A. Gardiner

# Infinite Processes
## Background to Analysis

With 182 Illustrations

Springer-Verlag
New York  Heidelberg  Berlin

A. Gardiner
Department of Mathematics
University of Birmingham
Birmingham, Warwickshire   B15 2TT
England

AMS Subject Classifications (1980): 00-01, 26-01, 26A03

Library of Congress Cataloging in Publication Data

Gardiner, A.
    Infinite processes, background to analysis.

    Includes index.
    1. Processes, Infinite.   I.  Title.
QA295.G315        519.5        81-8906 AACR2

Printed in the United States of America.

9 8 7 6 5 4 3 2 1

ISBN 0-387-90605-3 Springer-Verlag   New York  Heidelberg  Berlin
ISBN 3-540-90605-3 Springer-Verlag   Berlin  Heidelberg  New York

## Vorwort für den Kenner

*Dies Büchlein ist eine Konzession an die (leider in der Mehrzahl befindlichen) Kollegen, welche meinen Standpunkt...nicht teilen....*

*Die andere Richtung meint immer, während des späteren Verlaufes des Studiums würde der Schüler an Hand einer Vorlesung oder der Literatur die Sache schon lernen....*

*Wenn aber...dem einen oder anderen Kollegen der anderen Richtung die Sache so leicht erscheint, dass er sie in seinen Anfängervorlesungen (auf dem folgenden oder irgendeinem anderen Wege) bringt, würde ich ein Ziel erreicht haben, auf das ich in grösserem Umfange nicht zu hoffen wage.*

Edmund Landau, *Grundlagen der Analysis*

# Preface

As its title suggests, this book was conceived as a prologue to the study of *"Why the calculus works"*—otherwise known as *analysis*. It is in fact a critical reexamination of the infinite processes arising in *elementary* mathematics: Part II reexamines rational and irrational numbers, and their representation as infinite decimals; Part III examines our ideas of length, area, and volume; and Part IV examines the evolution of the modern function-concept.

The book may be used in a number of ways: firstly, as a genuine prologue to analysis; secondly, as a supplementary text within an analysis course, providing a source of elementary motivation, background and examples; thirdly, as a kind of postscript to elementary analysis—as in a senior undergraduate course designed to reinforce students' understanding of elementary analysis and of elementary mathematics by considering the mathematical and historical connections between them. But the contents of the book should be of interest to a much wider audience than this—including teachers, teachers in training, students in their last year at school, and others interested in mathematics.

The language and style adopted have both been kept as simple as is consistent with one of the book's main objectives: namely, to explain the need for *precision* when dealing with infinite processes, and to indicate as clearly as possible how this precision is achieved. Quite often a chapter is presented as an attempt to answer some specific question—chosen both for its intrinsic interest, and for the particular ideas which will emerge from our attempt to answer it. Each question is analysed in such a way as to emphasise those features of its solution which correspond to important ideas in analysis, but I have consistently resisted the temptation to believe that the process of understanding these ideas can be magically accelerated by confronting the reader with ready-made abstract formulations. As a result I hope that readers who do not go on to study analysis will still get a reasonable feeling for the careful analysis of certain specific infinite processes which arise in elementary mathematics. Readers who do go on to study analysis will meet abstraction quite soon enough. Last—but by no means least—I hope that both groups will enjoy working at some interesting elementary mathematics.

My sincere thanks go to those (in schools, teachers' colleges and universities) whose criticisms of the first draft challenged my own weary willingness to make do with second best. Just occasionally in life the best things do turn out to be free and totally unexpected. Among these many critics was one whom I have still never met—Tom Beldon. It was his original letter in the journal *Mathematics Teaching* (Number 86, 1979) which provoked me into writing; and his thoughtful, painstaking and persistent comments have kept me permanently on my toes.

> Die Feder kritzelt: Hölle das
> Bin ich verdammt zum Kritzelnmüssen?
> F. Nietzsche, *The Gay Science*

# Advice to the Reader

> In this way mankind stumbles on
> in its task of understanding the world.
>
> A. N. Whitehead, *The Function of Reason*

## The Text

The text is divided into four very different Parts.

Part I is short and largely descriptive. Using very broad brushstrokes, it paints a backcloth against which the action of Parts II, III and IV may be seen in some kind of perspective. You should not expect to understand everything here first time through; but if you can get the gist of the argument, the more important details should get reinforced as you work through the actual mathematics in Parts II, III, and IV. Perhaps Part I should be read a second time after you have worked through as much as you need from the rest of the book.

Part II is the longest section of the book. It examines in detail infinite processes arising in the realm of number. Part II begins with a slightly geometrical flavour, in that Chapter II.1 considers "numbers" as "ratios of (lengths of) line segments." Do not worry if your own geometrical experience is limited: simply extract what you can from Chapters II.1–II.4 in the knowledge that these ideas are not used again until Chapter II.13. You can then begin afresh with the most important bit of Part II, which is contained in Chapters II.5–II.12. In these central chapters we analyse the infinite processes which are *implicit* in the familiar representation of rational and irrational numbers as decimals. Most of the text is devoted to the analysis of specific examples. But these *specific* examples have been chosen and presented with two things in mind: firstly to indicate what is meant by *a careful and precise* analysis of infinite processes; and secondly to indicate as clearly as possible *how one should proceed in general.*

Part III explores the extent to which the familiar geometrical notions of length, area and volume depend on infinite processes. Because the context here is so much richer (considering, as it does, *numbers in geometry* rather than numbers all on their own), the exposition in Part III is neither so precise nor so complete as in Part II. The aim of Part III is to *begin* the process of analysing and taming our geometric intuition in order to get some feeling for what is involved, rather than to replace it altogether.

Modern mathematics is not so much the study of numbers and space as the study of *functions*, but one cannot hope to do justice to this fact in an elementary introduction such as this. In Part IV we outline very briefly some of the basic questions which result from the fact that naive calculus is based on the application of infinite processes not only to numbers and geometry, but also to *functions*. In particular we consider the crucial question: *What exactly is a function?* and take a closer look at the most familiar examples of functions (namely polynomial, rational, exponential and trigonometric functions).

## The Exercises

It may be necessary to stress that

> the Exercises are an integral part of the text.

Actively doing problems takes so much more time than passively reading the text, that it is always tempting to assume that exercises tucked away at the end of the chapter are a kind of optional appendix—a crutch for the feeble-minded, or a pastime for the unemployed. Nothing could be further from the truth. Most of the exercises in this book arise directly out of the text, and need to be understood and attempted in context. Thus, for example, in Chapter I.1 (on page 5) you will read

> ... there is no guarantee that, when $h$ is small, an infinite sum of terms involving $h$ will itself remain small (see Exercise 1, especially part (iii)).

You should turn immediately to Exercise 1 at the end of Chapter I.1 and check that you understand what the exercise is about, and why it is relevant to the discussion on page 5. Where possible you should work through the Exercise there and then rather than wait until you have finished reading the text of that particular chapter.

# Contents

## PART I
# FROM CALCULUS TO ANALYSIS

In which we discover
- how, around 1800, mathematicians began to realise that the lack of precision in their understanding and manipulation of the infinite processes involved in naive calculus was a source of error and confusion;
- that naive calculus therefore needs to be refashioned in a clear and precise way;
- that the need to revise our understanding of a piece of mathematics, such as naive calculus, should not really surprise us;
- but that the purely arithmetical way in which naive calculus was in fact refashioned (by around 1870) is slightly surprising.

# What's Wrong with the Calculus?

Obscurantism is the refusal to speculate freely
on the limitations of traditional methods.
A.N. Whitehead, *The Function of Reason*

The invention (around 1670) of the differential and integral calculus, its development, application and extension during the following two centuries, and the somewhat belated explanation (completed by about 1870) of *why* this calculus works, together constitute one of the major intellectual achievements of Western European culture.

The word *"calculus"* is in fact a general term describing any system of rules and symbols which effectively reduces the solution of a large class of mathematical, or logical, problems to some kind of routine calculation. The best known system of this kind is the differential and integral calculus, but it was by no means the first such calculus. For example, many numerical problems can be solved by simply applying the four standard processes for addition, subtraction, multiplication and division; the rules of algebra allow one to manipulate *variables* and *unknowns* as easily as numbers themselves; and coordinate geometry provides a routine way of solving many geometrical problems by means of elementary algebra. But however valuable these other calculi may be, they simply cannot compete with the power of the differential and integral calculus and the astonishing variety of important problems which it can help to solve. It is therefore hardly surprising that the more general meaning of the word calculus has been almost forgotten, and that the expression *the calculus* is now generally understood to refer to *the differential and integral calculus* alone.

For many students today, *the calculus* is the climax of their encounter with elementary mathematics: it exploits almost all the mathematics they have previously learnt, (especially algebra, trigonometry, graphs and coordinate geometry), and it opens the way to the analysis of a whole host of interesting problems (in elementary geometry, in differential geometry, in statics and dynamics, in physics—in fact, in any subject where we may reasonably assume that one quantity varies more or less *smoothly* with respect to some other quantity). The flexibility and power of the calculus is really very impressive and frequently provides students with their first glimpse of the potential intellectual fascination of mathematics.

"What then," you might reasonably ask, "is the meaning of the title of this chapter? Why go on about the impressive flexibility, power and intellectual fascination of the calculus when you promised to explain what is wrong with it? What, if anything, *is* wrong with it? And just supposing there is something wrong, how will the contents of this book help us to put it right?" Such questions are to be encouraged, even if they cannot be satisfactorily answered in a few pages: indeed the whole book is essentially a study of why one must and how one can introduce *precision* into certain infinite processes which are directly relevant to the task of sorting out the calculus. But though no brief explanation can be completely comprehensible to the beginner, I shall nevertheless indicate in this first introductory chapter how mathematicians gradually came to accept that their intuitive ideas about infinite processes in general, and about the calculus in particular, needed clarification. The second introductory chapter seeks to explain why this need to reshape the calculus should not really surprise us, what form the revised 1870 version of the calculus took, and why it took the form it did.

*Naive* calculus lacks one crucial ingredient—namely, *precision.* Its fundamental ideas depend on *intuitive* notions about infinite processes (such as taking limits and summing infinitely many terms) and about infinitesimal quantities. These intuitive ideas remain reasonably reliable as long as the problems tackled and the methods used are sufficiently elementary. But as soon as one begins, for example, to work with *multiply infinite processes* (such as taking the *limit*, as $h$ tends to 0, of an *infinite sum* of terms involving $h$) most people's intuition becomes decidedly unreliable. For example, to find the derivative of a simple-looking function[1] like $x^a$ we have to evaluate the limit, as $h$ tends to 0 of an expression involving $(x + h)^a$

$$\lim_{h \to 0} \frac{(x + h)^a - x^a}{h}$$

and we naturally use the binomial theorem to write $(x + h)^a$ as a sum of terms involving $h$. But if $a$ is anything other than a positive whole number or zero, then the binomial theorem expresses $(x + h)^a$ as an *infinite* sum; so

$$\frac{(x + h)^a - x^a}{h} = ax^{a-1} + \frac{a(a - 1)}{2!} x^{a-2}h + \frac{a(a - 1)(a - 2)}{3!} x^{a-3}h^2 + \cdots$$

We therefore find ourselves having to *take the limit*, as $h$ tends to 0, of an *infinite sum* of terms involving $h$. However, as we shall see in our discussion of the infinite sum

$$\frac{\sin h}{1} + \frac{\sin 3h}{3} + \frac{\sin 5h}{5} + \frac{\sin 7h}{7} + \cdots,$$

---

[1] In Chapter IV.3 we shall see that this function is not quite so simple as it looks.

there is no guarantee that, when $h$ is small, an infinite sum of terms involving $h$ will itself remain small (see Exercise 1, especially part (iii)).

It had in fact been recognised from the very beginning that the explanations as to *why* the calculus worked were inadequate and that their dependence on intuitive notions of infinitesimals and infinite processes was, from a strictly mathematical point of view, indefensible. Leibniz (1646–1716) suggested that the slope $dy/dx$ of the curve $y = f(x)$ was, in some sense, a genuine "quotient of two *infinitesimal quantities* $dy$ and $dx$"—$dx$ being an infinitesimally small change in $x$, and $dy$ the corresponding infinitesimally small change in $y$: that is,

$$dy = f(x + dx) - f(x).$$

But Leibniz never explained what kind of beast these *infinitesimal quantities* were; and whatever they were, they certainly behaved rather strangely at times. For example

if $y = u \cdot v = u(x) \cdot v(x)$
then $dy = u(x + dx) \cdot v(x + dx) - u(x) \cdot v(x)$.

But the infinitesimally small change $dx$ in $x$ causes not only an infinitesimally small change $dy$ in $y$, but also infinitesimally small changes $du$ in $u$ and $dv$ in $v$: thus,

$$du = u(x + dx) - u(x), \quad dv = v(x + dx) - v(x).$$

$$\therefore \ dy = d(u \cdot v) = (u + du)(v + dv) - u \cdot v$$

$$= u \cdot dv + du \cdot v + du \cdot dv.$$

Since Leibniz wished us to conclude that

$$dy = u \cdot dv + v \cdot du$$

it would appear that $du \cdot dv$ has to be equal to zero!

Newton (1642–1727) sought to avoid using infinitesimals, but his own explanations were far from adequate: (thus, for example, in the *Principia*—Book II, Lemma II—he gets around the difficulty of publicly discarding the term $du \cdot dv$ when differentiating a product $y = u \cdot v$ by employing a sleight-of-hand, which essentially involves claiming that, if a change $dx$ in $x$ gives rise to changes $du$ in $u$ and $dv$ in $v$, then the corresponding change $dy$ in $y$ is given by

$$dy = (u + \tfrac{1}{2}du) \cdot (v + \tfrac{1}{2}du) - (u - \tfrac{1}{2}du) \cdot (v - \tfrac{1}{2}dv)$$

$$= u \cdot dv + v \cdot du).$$

Newton tried various ways of justifying his "calculus," but one gets the distinct impression that he was not really satisfied with any of them.

But if the inadequacy of the usual explanations was recognised so early, why on earth did it take two hundred years to work out the first clear, non-contradictory explanation as to *why* the calculus works? I shall suggest two reasons.

Firstly, there was initially no obvious cause for concern! Perhaps the explanations as to why the calculus worked were weak, but it was obviously the right kind of theory to handle problems about tangents, areas and volumes, about shapes of curves in general, about centres of gravity and moments of inertia, and about speed, acceleration, and motion. And in each of these areas it delivered the goods!

In many ways beginning undergraduates are in the same sort of position as eighteenth century mathematicians: revelling in the joys that naive calculus can bring, and knowing little of its potential weaknesses. This can make it exceedingly difficult for such students to understand why one tries in analysis to reconstruct the calculus as carefully as possible. "Well," you might reply, "stop telling us we don't understand, and just show us some of these 'potential weaknesses' of the calculus. Surely we will then see that something needs to be done."

Which brings me to the second reason why it all took so long. There was never any sudden realisation that the naive view of the calculus was inadequate. A long chain of events gradually forced mathematicians to face up to certain rather subtle, but nonetheless real difficulties. One should not therefore expect to find nice elementary examples which will effectively convince beginners once and for all that we have no choice but to reconstruct the whole edifice of naive calculus on a more solid foundation. But rather than shirk the whole issue, let me try to give two examples of the way in which mathematicians came gradually to accept that something absolutely basic—namely, *precision*—was seriously lacking.

Firstly, where precision is lacking, one might not be surprised to find that different mathematicians who tackle the same problem in different ways may come up with different solutions. For example, the partial differential equation,

$$\frac{\partial^2 y}{\partial x^2} = \frac{1}{c^2}\frac{\partial^2 y}{\partial t^2}$$

represents the motion of a vibrating string which moves in the $x,y$-plane perpendicular to the $x$-axis. Solutions for the motion of such a vibrating string were obtained by D'Alembert (1717–1783) in 1747, and by Euler (1707–1783) in 1748; however, they gave their apparently similar solutions very different interpretations.[2] Far more serious was the fact that in 1753 Daniel Bernoulli (1700–1782) came up with a solution in the form of an infinite series of sines and cosines that looked totally different from the other two. But rival solutions, purporting to describe the behaviour of one

---

[2] For details consult the book *The Development of the Foundations of Mathematical Analysis from Euler to Riemann*, by I. Grattan Guinness (published by MIT, 1970), pages 2–9.

and the same vibrating string, generate a tension which cannot be simply ignored, and which must somehow be resolved.

I shall describe the second example, which also concerns infinite series of functions, in greater detail. To raise certain important issues let us consider the usual kind of naive approach to finding a Taylor series expansion for the function $f(x)$. One probably begins by simply assuming that the function $f(x)$ can be written as an infinite power series:

$$f(x) = a_0 + a_1 x + a_2 x^2 + a_3 x^3 + \cdots.$$

We then substitute $x = 0$ on both sides to get $f(0) = a_0$. Next we differentiate both sides

$$f'(x) = a_1 + 2a_2 x + 3a_3 x^2 + \cdots.$$

Substituting $x = 0$ on both sides then shows that $f'(0) = a_1$. Continuing in this way, we show that the $n^{\text{th}}$ coefficient $a_n = f^{(n)}(0)/n!$ : hence

$$f(x) = \sum_{n=0}^{\infty} \frac{f^{(n)}(0)}{n!} x^n.$$

But what right had we to assume that $f(x)$ could be expressed as an infinite power series in the first place? And can one just substitute $x = 0$ in an *infinite* power series as though it were an ordinary polynomial? And have we not made rather a large jump in assuming that, just because the sum (say $x^2 + \cos x$) of *two* functions ($x^2$ and $\cos x$) can be differentiated term by term, the sum

$$a_0 + a_1 x + a_2 x^2 + a_3 x^3 + \cdots$$

of *infinitely many* functions can therefore also be differentiated term by term?

Some of the things we simply take for granted in naive calculus turn out to be more or less correct; but we should not be surprised to discover that occasionally, something which we have simply taken for granted, turns out to be false. For example, the partial differential equation

$$K \cdot \frac{\partial^2 v}{\partial x^2} = \frac{\partial v}{\partial t}$$

is called the "heat equation," because it represents the way in which heat diffuses along a metal bar (the bar is assumed to be lying along the $x$-axis, and $v = v(x,t)$ is the temperature of the point $x$ at the time $t$). Joseph Fourier (1768–1830) discovered a method of pioducing what he believed to be the most general solution of the heat equation. His method seemed to show that, provided we are only interested in the values of $x$ which lie in a particular interval (say $a \leq x \leq b$), then *every function $f(x)$ can be expressed as an infinite sum of sines and cosines.* For example, for $-\pi \leq x \leq \pi$, the function

$$f(x) = \begin{cases} -1 & \text{if} & -\pi < x < 0, \\ 0 & \text{if} & x = -\pi, 0, \text{ or } \pi, \\ 1 & \text{if} & 0 < x < \pi, \end{cases}$$

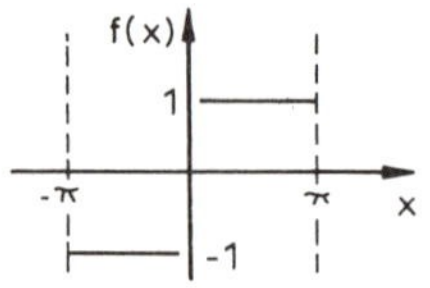

Figure 1

whose graph is shown in Figure 1, can be expressed (see Exercise 1(ii)) as an infinite sum of sines:

$$f(x) = \frac{4}{\pi}\left(\frac{\sin x}{1} + \frac{\sin 3x}{3} + \frac{\sin 5x}{5} + \cdots\right).$$

Fourier tried to publish his results in 1807, but their publication was blocked. Among the many objections of his contemporaries was their persistent belief that

   since the functions $(\sin x)/1$, $(\sin 3x)/3$, $(\sin 5x)/5$, ... all have nice continuous graphs, and

   since any *finite sum* of such functions also has a nice continuous graph,

   the *infinite sum* $(\sin x)/1 + (\sin 3x)/3 + (\sin 5x)/5 + \ldots$ must also have a nice continuous graph, and so could not possibly be equal to a function like $f(x)$, which has a jump in its graph at $x = 0$.

The calculus comes in here not chiefly because Fourier was trying to solve a *differential* equation, but because of the way he used *integration* to calculate the coefficients in his infinite series of sines and cosines (Exercise 2).

   You should not expect to understand these examples completely, but once you have had a go at the exercises you should at least begin to appreciate the potential pitfalls of continuing to treat infinite processes in a naive way.

   The investigation of the historical background, of what exactly was wrong with the explanations of the calculus current throughout the eighteenth century, of the unsuccessful efforts of various eighteenth century mathematicians to do better, of the developments (around 1800) which made mathematicians much more conscious that there was a genuine problem, of the way in which it was originally resolved (between 1820 and 1870), and of how analysis developed thereafter, would make a most instructive study.[3] But the aim of this book is much more modest.

---

[3]For an introduction to the ideas *leading up to* the invention of the calculus, see O. Toeplitz: *The Calculus: A Genetic Approach* (University of Chicago Press, 1963). A recent book edited by I. Grattan Guinness, *From the Calculus to Set Theory, 1630–1910* (Duckworth, 1980), covers the whole period. A more elementary and selective, but still valuable, introduction to the middle period (1750–1820) may be obtained from the first few chapters of I. Grattan Guinness, *The Development of the Foundations of Mathematical Analysis from Euler to Riemann* (M.I.T., 1970).

This is not a history of the calculus, but rather a collection of material addressed to the reader who is familiar with the elementary ideas behind the differential calculus and who is prepared to use the material to reflect on, and possibly to revise, some of his or her own mathematical ideas—all with a view to appreciating the kind of difficulties which had to be overcome before the calculus could be given a solid *arithmetical* foundation. The ideas we shall need to reflect on include

> the existence of irrational numbers,
> the nature of real numbers,
> infinite decimals,
> the nature of infinite processes,
> the idea of a function,

and, perhaps above all,

> the relationship between calculus and geometry, and
> the reasons for insisting on their separation.

The material we shall explore concerns some of the important elementary examples of infinite processes which you will have met in school, but which you have probably never investigated too closely (infinite decimals; length, area and volume; exponential and trigonometric functions). These examples are of considerable interest, but they are not explored here merely for their own sake: we are far more interested in the light they can shed on *the subtlety and the significance of the jump from finite to infinite processes* and on *the way in which infinite processes can be safely treated mathematically.*

## EXERCISES

1.  (i) Make reasonably accurate sketches of the graphs of the following functions:

(a) $y = \dfrac{\sin x}{1}$, $\quad y = \dfrac{\sin 3x}{3}$, $\quad y = \dfrac{\sin 5x}{5}$, $\quad y = \dfrac{\sin 7x}{7}$, $\ldots$

(b) $y = \dfrac{\sin x}{1} + \dfrac{\sin 3x}{3}$, $\quad y = \dfrac{\sin x}{1} + \dfrac{\sin 3x}{3} + \dfrac{\sin 5x}{5}$,

$y = \dfrac{\sin x}{1} + \dfrac{\sin 3x}{3} + \dfrac{\sin 5x}{5} + \dfrac{\sin 7x}{7}$, $\ldots$

(ii) Investigate at least the first twenty terms of the sequence

$\dfrac{\sin x}{1}$, $\quad \dfrac{\sin x}{1} + \dfrac{\sin 3x}{3}$, $\quad \dfrac{\sin x}{1} + \dfrac{\sin 3x}{3} + \dfrac{\sin 5x}{5}$,

$\dfrac{\sin x}{1} + \dfrac{\sin 3x}{3} + \dfrac{\sin 5x}{5} + \dfrac{\sin 7x}{7}$, $\ldots$

for each of the following values of $x$ (*in radians*):

(a) $x = 1$,   (b) $x = .1$,   (c) $x = .01$;      (d) $x = 0$;
(e) $x = -.1$,   (g) $x = -.01$.

Explain the pattern that emerges. Do your results support the assertion made on page 8 that the infinite series

$$\frac{\sin x}{1} + \frac{\sin 3x}{3} + \frac{\sin 5x}{5} + \frac{\sin 7x}{7} + \frac{\sin 9x}{9} + \cdots$$

takes the value 0 when $x = k\pi$, and the value $\pi/4$ for all other values of $x$?

[This is an excellent opportunity to introduce yourself to a programmable calculator: the program is very simple and the effect of seeing the series summed term by term is quite dramatic. You need two memories (one to hold the sequence of numbers 1, 3, 5, 7, ... in turn and the other to hold the sequence of partial sums) and a GOTO button. Choose the value of $x$—say 1 *radian*; then program the following crude, but very simple sequence of instructions:

(1) put the number 1 in the first memory M1;
(2) put the number 0 in the second memory M2;
(3) "label": recall M1, multiply by $x$, take the sine, divide by M1, and add this to the number in M2;
(4) recall M2 and "Pause" [to allow the partial sum calculated so far to become visible in the display];
(5) recall M1 and "Pause" [to allow yourself to see what value of $2n + 1$ has been reached];
(6) add 2 to the number in M1;
(7) GOTO "label".]

(iii) The honest ones amongst you will probably admit that you make no attempt to evaluate *the limit as h tends to zero* when you calculate

$$\lim_{h \to 0} \frac{(x+h)^a - x^a}{h} = \lim_{h \to 0} \left( \frac{x^a + ax^{a-1}h + [a(a-1)/2!]x^{a-2}h^2 + \cdots - x^a}{h} \right)$$

$$= \lim_{h \to 0} \left( ax^{a-1} + \frac{a(a-1)}{2!} x^{a-2}h + \frac{a(a-1)(a-2)}{3!} x^{a-3}h^2 + \cdots \right)$$

$$= ax^{a-1}$$

Instead you simply *substitute h = 0* in the expression

$$ax^{a-1} + \frac{a(a-1)}{2!} x^{a-2}h + \frac{a(a-1)(a-2)}{3!} x^{a-3}h^2 + \cdots$$

*which is the one thing the notion of "limit" forbids!* Suppose we write

$$h \to 0^+ \qquad \text{and} \qquad h \to 0^-$$

to mean

   "$h$ tends to zero from above"   and   "$h$ tends to zero from below,"

respectively.

What does part (ii) suggest about the values of

$$\lim_{h \to 0^+} \left( \frac{\sin h}{1} + \frac{\sin 3h}{3} + \frac{\sin 5h}{5} + \frac{\sin 7h}{7} + \cdots \right)$$

and

$$\lim_{h \to 0^-} \left( \frac{\sin h}{1} + \frac{\sin 3h}{3} + \frac{\sin 5h}{5} + \frac{\sin 7h}{7} + \cdots \right) ?$$

2. Let

$$f(x) = \begin{cases} -1 & \text{if} \quad -\pi < x < 0 \\ \phantom{-}0 & \text{if} \quad x = -\pi, \, 0, \text{ or } \pi \\ \phantom{-}1 & \text{if} \quad 0 < x < \pi \end{cases}$$

Suppose that

$$f(x) = \tfrac{1}{2}a_0 + \sum_{n=1}^{\infty} (a_n \cos nx + b_n \sin nx)$$

(i) Show that

$$\int_{-\pi}^{\pi} \sin mx \sin nx \, dx = \begin{cases} 0 & \text{if } m \neq n \\ \pi & \text{if } m = n \neq 0 \end{cases}$$

$$\int_{-\pi}^{\pi} \sin mx \cos nx \, dx = 0$$

$$\int_{-\pi}^{\pi} \cos mx \cos nx \, dx = \begin{cases} 0 & \text{if } m \neq n \\ \pi & \text{if } m = n \neq 0 \end{cases}$$

(ii) Use the fact that $f(-x) = -f(x)$ to show that all the coefficients $a_n$ are equal to zero.

(iii) We now have

$$f(x) = \sum_{n=1}^{\infty} b_n \sin nx$$

For a fixed value of $m$, multiply both sides by $\sin mx$ and integrate between $-\pi$ and $+\pi$. This gives

$$\int_{-\pi}^{\pi} f(x) \sin mx \, dx = \int_{-\pi}^{\pi} \left( \sum_{n=1}^{\infty} b_n \sin nx \sin mx \right) dx.$$

*If you now make the optimistic assumption that we can safely integrate the infinite sum*

$$\sum_{n=1}^{\infty} b_n \sin nx \sin mx$$

*"term by term" as you would a finite sum*, then you should be able to show

that

$$b_m = \frac{1}{\pi} \int_{-\pi}^{\pi} f(x) \sin mx \, dx = \frac{2}{\pi} \int_0^{\pi} \sin mx \, dx = \begin{cases} 0 & \text{if } m \text{ is even} \\ 4/\pi m & \text{if } m \text{ is odd} \end{cases}$$

Thus

$$f(x) = \frac{4}{\pi} \left( \frac{\sin x}{1} + \frac{\sin 3x}{3} + \frac{\sin 5x}{5} + \frac{\sin 7x}{7} + \cdots \right).$$

(iv) Find the value of the infinite sum

$$\sum_{r=0}^{\infty} \frac{(-1)^r}{2r+1}$$

3. You may have noticed that the infinite series

$$\sum_{r=0}^{\infty} \frac{\sin(2r+1)x}{2r+1}$$

converges only very slowly towards $\pi/4$ (provided of course $x \neq k\pi$). The reason is simply that after 500 terms, each term may still be contributing almost 1/1000 to the running total, and these contributions all mount up to produce the kind of oscillating behaviour you should have observed. Investigate the behaviour of the following infinite series:

$$\text{(a)} \sum_{r=0}^{\infty} \frac{1}{2r+1} \, ; \quad \text{(b)} \sum_{p \text{ prime}} \frac{1}{p} \, ; \quad \text{(c)} \sum_{r=1}^{\infty} \frac{1}{r^2} \, .$$

# Growth and Change in Mathematics

> In the study of ideas, it is necessary to remember, that
> insistence on hard-headed clarity issues from sentimental
> feeling, as it were a mist, cloaking the perplexities of fact.
> Insistence on clarity at all costs is based on sheer
> superstition as to the mode in which human intelligence
> functions. Our reasonings grasp at straws for premises and
> float on gossamers for deductions.
>
> A.N. Whitehead, *Adventures of Ideas*

You may notice that the title (and the contents) of this and the previous chapter contradict the idea that mathematics consists of a fixed and un-challengeable stock of truths. Should this come as a surprise? How are mathematical ideas born? How do they grow? In what sense does math-ematics itself "evolve"? There are as one might expect no easy answers. In the long term one simply has to keep such questions permanently in the back of one's mind, ready to take advantage of any new example which might provide some unexpected insight. But in the short term, the challenge is to begin, somehow or other, to make sense of such questions. We should not perhaps expect at the outset to make much sense of *historical* examples, but we can at least begin by reflecting how *our own* view of mathematics has changed. If we do this, then it should soon become apparent that our own private view of mathematics is never rigidly fixed, but changes as we grow. Some of these changes are of a *local* nature, in that they concern *specific* problems or ideas: for example, there was a time when each of us really believed

5 *take away* 8 *can't be done*   and   16 *into* 12 *won't go,*

though we are now quite happy to write

$$5 - 8 = -3, \quad \text{and} \quad 12 \div 16 = \tfrac{3}{4} = 0.75.$$

Other changes affect our *global* perspective, in the sense that they lead us to revise our perception of the relative importance of different parts of math-ematics, and our awareness of connections between them. For example, there was a time when we knew nothing of algebra and trigonometry, or of tangents and exponents, and when arithmetic ruled supreme. Later, we studied algebra, geometry, tangents, areas, exponents and trigonometry

13

separately, but we knew nothing about fundamental links between them. Later still we came to realise that algebra and geometry together make co-ordinate geometry possible, that tangents and areas correspond to derivatives and integrals in the calculus, and that exponential and trigonometric functions are indissolubly linked by the relation

$$e^{i\theta} = \cos\,\theta + i\,\sin\,\theta.$$

Once we begin to realise just how much *our own view* of mathematics has changed over the years, it should come as no surprise to learn that *mathematics itself* is also constantly changing. But how does one square this fact with the popular view of mathematics as a *fixed and unchallengeable stock of truths*? There is, unfortunately, an all too easy way out of this dilemma. The idea that mathematics is constantly changing does not represent a very serious challenge to the popular view provided the changes which occur can be interpreted as the *progressive extension and improvement* of existing ideas and methods: for mathematics can then be seen as a *progressively expanding*, but still *unchallengeable*, stock of truths. The idea that historical change in mathematics is *progressive* is especially attractive to those who interpret the adjustments we have to make in order to accept statements like

$$5 - 8 = -3, \quad \text{and} \quad 12 \div 16 = \tfrac{3}{4} = 0.75$$

as natural and very welcome extensions of ideas which we had previously understood only in part.

But the view that historical change in mathematics is *progressive* is simply false: mathematics has a much more turbulent, messy, and interesting history than one might expect. It is true that many changes do emerge more or less smoothly as a result of the application, extension and improvement of existing ideas and methods. But many other changes are decidedly ragged and painful—with periods of relative security flowing into periods of darkness and confusion, whose possible resolution is often "sensed" rather than discovered, perhaps on the basis of very flimsy evidence (see Exercise 2). Some of these untidy, but exciting, insights come to nothing. Others are premature—lacking either clarity or any conceivable pay-off (as was the case with negative and, to a lesser extent, complex numbers, which were eventually accepted several centuries after they were first proposed). Others still are soon clarified, are widely used, and quickly become entirely respectable (as was the case with logarithms). But some turn out to involve hidden mathematical subtleties which are very difficult to explain, *though the methods are recognised as being very important and are widely used* (as was the case with the calculus). Mathematics, it would appear, does not grow by smooth and steady progress alone.

And on reflection, many of us would have to admit that *our own* mathematical development has not always been as straightforward as the revision of the childhood beliefs

> 5 *take away* 8 *can't be done*   and   16 *into* 12 *won't go.*

For example, most of us learn to *add* (positive and) *negative* numbers together more or less as we add *positive* whole numbers—perhaps by shifting to the right and to the left on a mental "number line" or "axis." But the problem of *multiplying negative* numbers is often reduced to following some rule, and cannot be easily interpreted in terms of the mental pictures we use for multiplying *positive* whole numbers $a,b$—such as "*a lots of b things.*" The reason for this conceptual discontinuity is simply that most answers to the question

> "*How should one think of negative numbers?*"

are based on ideas of *addition* (and subtraction) and do not help us to think meaningfully about the *multiplication* of such numbers.

Some of those who surmount this particular hurdle more or less comfortably, have difficulty later on when they meet complex numbers. The natural geometrical interpretation of multiplication by the complex number $z$ (namely, stretch by $|z|$ and rotate through arg $z$) may help to resolve any residual misgivings about the multiplication of *negative* numbers, but *complex* numbers introduce perceptual problems of their own.

*Counting* numbers arise first as a way of *counting*, or *measuring*, discrete collections of objects. *Real* numbers arise in scientific and everyday measurement, in co-ordinate geometry and the arithmetic of solving equations. In each of these contexts the real numbers we use appear to be a meaningful extension of the original notion of counting, or measuring, number; (admittedly we do use the *geometrical* number line as an aid to thought, but it is sufficiently like a ruler not to undermine the idea that we are working with genuine measuring numbers). In disturbing contrast, *complex* numbers can appear rather artificial—emerging as they do from a kind of generalised algebra, rather than as a natural extension of real measuring numbers. That we are prepared to accept complex numbers at all depends largely on their geometrical representation as *points in the plane*—a fact which serves only to reinforce our feelings about their artificiality *as numbers*.

Those who have no difficulty accepting negative and complex numbers are bound to find sticking points elsewhere: for learning mathematics involves one in an endless sequence of revisions of ideas one had previously come to accept as "final." Sometimes the necessary revision is a relatively mild affair, requiring only some *slight adjustment*, adaptation, or extension of former ideas—as some people find with the introduction and addition of negative numbers. On other occasions further progress in mathematics requires us to make some *radical reinterpretation* of old ideas—as is the case when someone who has always thought of the multiplication of "*a by b*" in terms of "*a* lots of *b* things" is confronted with the problem of multiplying *negative* numbers.

Three things should be stressed at this point. Firstly, one man's *slight adjustment* is another man's *radical reinterpretation*—since the extent of the adjustment one has to make depends on the flexibility of one's previous viewpoint. Secondly, even when a student, or a professional mathematician, is aware of the need for some radical reinterpretation, she or he may not in fact make the required change—either because the expected pay-off does not seem to justify the effort involved in making the change, or because she or he simply does not know what "reinterpretation" to make. And finally, the need for a radical reinterpretation can always be temporarily circumvented by reducing everything to unexplained rules (like "*two minutes make a plus*", or "$\sqrt{-1}$ *is that number whose square is* $-1$" or

$$\text{``}\frac{d}{dx}(x^a) = ax^{a-1}\text{''}).$$

Clearly, *too many* unexplained rules overload one's memory and undermine one's understanding; but when a method which obviously works has no satisfactory explanation, a rough justification accompanied by a good notation and a clear set of rules serves almost as well—at least for a time. One classic example of this phenomenon is Leibniz' *differential calculus*.

Leibniz presented his calculus in terms of the plausible, but unexplained, idea of allowing a variable $x$ to change by an *infinitesimal* quantity $dx$. He was certainly in no position to give either a precise meaning to the idea of an infinitesimal quantity, or a justification for the rules whereby they were manipulated; but the jiggery-pokery which resulted from the application of his unexplained rules was enormously fruitful; and his marvellously suggestive notation of "differentials"

$$\left(\text{as in } \frac{du}{dv}\cdot\frac{dv}{dx}=\frac{du}{dx}, \qquad d(u\cdot v)=u\cdot dv+v\cdot du, \qquad \text{etc.}\right)$$

is still very much with us.

Ever since the time of the ancient Greeks, mathematicians and philosophers had been fascinated by the infinitely large and the infinitesimally small. The sixteenth century witnessed a revival of interest in the idea that matter is composed of very small *indivisible atoms*—an idea which was originally attributed to the Greek Democritus (born around 460 B.C.). At the same time, newly rediscovered works on the calculation of surface areas, volumes and centres of gravity, which had been written by Archimedes (who was killed in 212 B.C. at the siege of Syracuse), were printed in Italy and circulated throughout Europe. These inspired Stevin (1548?–1620?), Kepler (1571–1630), Galileo (1564–1642) and others to develop plausible ways of using infinitesimally small quantities to obtain answers to questions in geometry, in dynamics, and in statics. But their ad hoc methods were unreliable, and produced nonsense almost as often as sense.

*Leibniz's rules*[1] *tamed the infinitesimally small so successfully, that even the ordinary user could harness its potential with relatively little fear of going astray.* But no matter how fruitful they may have been, Leibniz's rules remained *heuristic devices*—brilliant *rules of thumb*—whose mathematical justification was unknown.

The precise reason why these rules worked continued to elude mathematicians throughout the eighteenth century; but there were so many wonderful things that one could do with the aid of (naive) calculus, that it seemed petty to complain about this lack of a solid mathematical foundation. Thus, for a long time, those working with infinite processes remained largely dependent on their intuitions and their experience. But in the course of the eighteenth and early nineteenth centuries those intuitions were repeatedly found wanting, and it gradually became clear that the fundamental principles on which the calculus was based could no longer be simply overlooked. The period after 1670 had seen the complete triumph of Leibniz's infinitesimal calculus over all rival methods: the English held out for a time against this foreign usurper (Leibniz), but by 1800 even they had begun to admit that Leibniz's methods and notation were far more flexible than Newton's. Thus it was the infinitesimal calculus of Leibniz, rather than Newton's fluxional calculus, which had to be reconstructed in the course of the nineteenth century.

Any successful attempt to present mathematics in an orderly way achieves the appearance of orderliness by excluding awkward material, and by deliberately selecting one particular approach to the exclusion of the many conceivable alternatives. The reconstruction of the infinitesimal calculus, which was completed around 1870, is no exception. But the particular approach used, and the alternatives which were excluded, were not at all what one might have expected—as I shall now explain.

During the first half of the nineteenth century, the most influential textbooks presented the infinitesimal calculus either as a part of *geometry* or as a new kind of *symbolical algebra* (that is, the manipulation of abstract symbols, like *dx*, according to specified rules). However the 1870 version of the calculus was *not* based on a clarified notion of infinitesimal, and *deliberately avoided* presenting the calculus as a branch of either geometry or

---

[1]Any attempt to understand how mathematics evolves must come to terms with the fact that Leibniz's work on infinitesimals, indivisibles, infinity and continuity was, in his own view, simply one aspect of his attempt to build a universal philosophical system which would illuminate the whole of reality. Nor is Leibniz an isolated example of this phenomenon. Descartes' coordinate geometry—without which neither Newton nor Leibniz could have begun to represent rates of change as slopes of tangents, or to relate tangents to areas—arose in a similar way: his work on geometry was published in 1637 as one of three appendices to his book, *Discourse on the Method of Rightly Conducting the Reason, and Seeking Truth in the Sciences*, and was apparently offered as evidence to support the claim that his general *method* could be employed to clarify mathematics as well as scientific and philosophical questions.

symbolical algebra[2]; instead its preferred approach was to develop the whole of elementary calculus *on the basis of the arithmetic of ordinary whole numbers alone.*

What then were the *general* reasons for divorcing the calculus from both geometry and symbolical algebra? One reason was that between 1800 and 1870 a whole host of different geometries and symbolical algebras had sprung up in a totally unexpected way. A more critical analysis of Euclid's geometry had revealed previously unidentified weaknesses in its logical structure; and this, together with the discovery of all sorts of non-euclidean geometries, completely devastated the classical view that *geometry offered a uniquely dependable foundation for the mathematics of continuous magnitudes.* Thus, though geometrical examples were, and still are, important and useful for the beginner, the calculus could never be sorted out by appealing to geometrical ideas—at least, not until geometry had itself been sorted out to some extent (something which was achieved only *as a result* of the reworking of the ideas behind the 1870 version of the calculus). Moreover, though symbolical algebra had proved to be a very powerful means of describing and analysing *familiar* processes (as with George Boole's, *Investigation of the Laws of Thought*, published in 1854), it could offer no *independent* explanation of the meaning of the symbols, nor of the rules according to which the symbols were manipulated, and so could never help remove the uncertainty surrounding the foundation of the calculus.

And what about infinitesimals and differentials? Though previous attempts to clarify their mathematical basis had been unsuccessful, the undeniable effectiveness of infinitesimals and differentials in the calculus naturally lead one to suspect that they must necessarily correspond to some solid mathematical entities. Such a suspicion is now known to be entirely justified[3]: a precise theory of infinitesimal quantities was worked out in the 1960's. But this was simply not available in the nineteenth century. Anyway, infinitesimals are also more subtle than ordinary real numbers, and are thus in some sense less suitable as a foundation on which to build the simplest, cleaned-up version of the calculus. But in mathematics, as elsewhere, simplicity and precision do not come easily. Euclid's *Elements* (ca. 300 B.C.) presented a cleaned-up version of ancient Greek mathemat-

---

[2]Cauchy's *Cours d'Analyse* (1821) is in part an attempt to free analysis from these twin millstones. In the introduction Cauchy points out that an algebraical formula "remains true only under certain conditions, and for certain values of the quantities to which it refers": he therefore rejects the practice of appealing to universal algebraical laws to justify calculations (e.g. with infinite, and even divergent, series). Cauchy also rejects the "grave error of thinking that one finds certitude only in geometrical proofs."

[3]A marvellous introduction to infinitely small and infinitely large numbers may be found in Donald Knuth's lovely little book, *Surreal Numbers: How Two Ex-Students Turned on to Pure Mathematics and Found Total Happiness, a Mathematical Novelette* (Addison-Wesley, 1974). An idea of how such numbers can be used in analysis may be obtained, for example, from the book, *Infinitesimal Calculus* by Henle and Kleinberg (MIT, 1979).

ics on the basis of the simplest foundation available at the time—namely a geometry of points and lines, which were presumed to satisfy only the very simplest conditions; but the deliberate choice of this simple *geometric* framework made it exceedingly awkward for Euclid to handle what we would interpret as elementary properties of *arithmetic* and *algebra*. The same is to some extent true of the first cleaned-up version of the calculus: to be able to build the calculus from *the arithmetic of whole numbers* alone is both surprising and aesthetically satisfying; but the methods which we have to invent in order to achieve this goal, can appear rather involved. They are the fruits of centuries of brilliant insights and patient sifting, and hold the key to understanding the calculus. They deserve serious study. But they are by no means the final word.

EXERCISES

The mathematician in you may agree that there is something unsatisfactory about merely calculating some kind of decimal approximation for infinite sums, especially when the infinite sums are, in some sense, as "natural" as

$$\sum_{r=0}^{\infty} \frac{(-1)^r}{2r+1}, \qquad \sum_{r=1}^{\infty} \frac{1}{r^2}, \qquad \text{or} \qquad \sum_{r=1}^{\infty} \frac{(-1)^{r+1}}{r}.$$

If you used a calculator to calculate partial sums for the *first* of these infinite sums (which is just $(\sin x)/1 + (\sin 3x)/3 + (\sin 5x)/5 + \cdots$ with $x = \pi/2$), then you might eventually decide that the first three decimal places of the sum appeared to be .785 ..., but you would hardly guess that the exact value was $\pi/4$! In these exercises we shall see the kind of brilliantly imaginative guesswork which allowed Leonhard Euler[4] to discover the exact value of the second infinite sum $\sum_{r=1}^{\infty} (1/r^2)$. The example serves a dual role: it is in part a demonstration of the extent to which mathematics evolves through *creative leaps*, and not just in a tidily logical way; but it is also an indication of the way in which the very success of these creative leaps raises serious questions for mathematics as a whole.

1. Show that, for each value of $N$,

$$1 < \sum_{r=1}^{N} \frac{1}{r^2} < 2.$$

Calculate $\sum_{r=1}^{N} (1/r^2)$ for $N = 10, 20, 30, 40, 50$.
(You would have to sum the first 1000 or so terms to get the first three decimal places of the infinite sum.)

2. We shall now consider the method whereby Euler discovered the exact value of the infinite sum $\sum_{r=1}^{\infty} (1/r^2)$. It would be sacrilege to interrupt the flow of Euler's argument with petty moralising sermons about the pitfalls which his genius miraculously avoids in this case. However, lesser mortals need to reflect on these

---

[4]This and many other beautiful examples of the role of systematic imaginative guesswork in mathematics can be found in the book *Induction and Analogy in Mathematics* (Oxford University Press, 1954) by George Pólya.

pitfalls, and for them my moralising footnotes (1)–(7) are collected at the end of this exercise.

Glattes Eis
Ein Paradeis
Für den, der gut zu tanzen weiss.

F.Nietzsche, *The Gay Science*

Euler begins by observing that, provided we allow complex roots, any ordinary polynomial

$$p(x) = a_0 + a_1 x + a_2 x^2 + \cdots + a_{n-1} x^{n-1} + x^n$$

can be factorised as the product of linear factors[1]—one for each of its $n$ roots $\alpha_1$, $\alpha_2, \ldots, \alpha_n$ :

$$p(x) = (x - \alpha_1)(x - \alpha_2)(x - \alpha_3) \cdots (x - \alpha_n).$$

These two expressions for $p(x)$ tell us that the coefficient $a_{n-1}$ of $x^{n-1}$ is given by a *finite sum*

$$a_{n-1} = -(\alpha_1 + \alpha_2 + \alpha_3 + \cdots + \alpha_n).$$

Euler's idea here, as in many of his brilliant discoveries, is to *let $n \to \infty$ in an intelligent way*, and so find some way of using the relation

$$a_{n-1} = -(\alpha_1 + \alpha_2 + \cdots + \alpha_n)$$

to calculate *infinite sums*.

At first sight it looks as though this cannot be done: if we let $n \to \infty$ in the obvious way, then we get an *endless* polynomial[2]

$$P(x) = a_0 + a_1 x + a_2 x^2 + a_3 x^3 + a_4 x^4 + \cdots$$

and even supposing we manage to "factorise" this as an "endless product"[3]

$$P(x) = (x - \alpha_1)(x - \alpha_2)(x - \alpha_3) \cdots$$

the infinite sum

$$-(\alpha_1 + \alpha_2 + \alpha_3 + \alpha_4 + \cdots)$$

would not be the coefficient of anything at all!

So Euler reinterprets the factorisation of ordinary (finite) polynomials in a way which at least has a chance of giving us what we want:

Any ordinary polynomial

$$p(x) = a_0 + a_1 x + a_2 x^2 + a_3 x^3 + \cdots + a_{n-1} x^{n-1} + x^n$$

in which $a_0 \neq 0$, can be factorised as a product of $n$ linear factors—one for each of its $n$ roots $\alpha_1, \alpha_2, \alpha_3, \ldots, \alpha_n$ :

$$p(x) = a_0 \cdot \left(1 - \frac{x}{\alpha_1}\right)\left(1 - \frac{x}{\alpha_2}\right)\left(1 - \frac{x}{\alpha_3}\right) \cdots \left(1 - \frac{x}{\alpha_n}\right)^{[4]}.$$

In this form it is the coefficient $a_1$ of $x$ which appears as a *finite sum*

$$a_1 = -a_0 \cdot \left(\frac{1}{\alpha_1} + \frac{1}{\alpha_2} + \frac{1}{\alpha_3} + \cdots + \frac{1}{\alpha_n}\right).$$

(Moreover, this sum of reciprocals looks much more likely to help us calculate $\sum (1/r^2)$.)

If we now let $n \to \infty$ then we get an "endless polynomial"

$$P(x) = a_0 + a_1 x + a_2 x^2 + a_3 x^3 + a_4 x^4 + \cdots$$

and *provided* $a_0 \neq 0$, we might hope to be able to factorise this as an "endless product" of linear factors—one for each of the "roots" $\alpha_1, \alpha_2, \alpha_3, \ldots$ of $P(x)^{(5)}$:

$$P(x) = a_0 \cdot \left(1 - \frac{x}{\alpha_1}\right)\left(1 - \frac{x}{\alpha_2}\right)\left(1 - \frac{x}{\alpha_3}\right) \cdots$$

If we could do all this, then we might expect that the coefficient $a_1$ of $x$ in $P(x)$ would satisfy

$$a_1 = -a_0 \cdot \left(\frac{1}{\alpha_1} + \frac{1}{\alpha_2} + \frac{1}{\alpha_3} + \frac{1}{\alpha_4} + \cdots\right).$$

Our task is therefore to find an "endless polynomial" whose "roots" $\alpha_1, \alpha_2, \alpha_3, \alpha_4, \ldots$ allow us to calculate the infinite sum

$$\frac{1}{1^2} + \frac{1}{2^2} + \frac{1}{3^2} + \frac{1}{4^2} + \cdots$$

Here Euler takes the familiar function $\sin x$, which can be written as an "endless polynomial"

$$\sin x = x - \frac{x^3}{3!} + \frac{x^5}{5!} - \frac{x^7}{7!} + \frac{x^9}{9!} - \frac{x^{11}}{11!} + \cdots$$

and which has "roots"[6]

$$0, \quad \pm\pi, \quad \pm 2\pi, \quad \pm 3\pi, \quad \pm 4\pi, \quad \pm 5\pi, \cdots$$

Unfortunately the constant term $a_0 = 0$, so we divide by $x$ to get an endless polynomial for $\sin x / x$

$$\frac{\sin x}{x} = 1 - \frac{x^2}{3!} + \frac{x^4}{5!} - \frac{x^6}{7!} + \frac{x^8}{9!} - \frac{x^{10}}{11!} + \cdots$$

which has "roots"

$$\pm\pi, \quad \pm 2\pi, \quad \pm 3\pi, \quad \pm 4\pi, \quad \pm 5\pi, \cdots$$

If we are very fortunate, then we might find that the "endless polynomial" for $\sin x / x$ factorises as an "endless product" of linear factors:

$$\frac{\sin x}{x} = 1 \cdot \left(1 - \frac{x}{\pi}\right)\left(1 + \frac{x}{\pi}\right)\left(1 - \frac{x}{2\pi}\right)\left(1 + \frac{x}{2\pi}\right)\left(1 - \frac{x}{3\pi}\right)\left(1 + \frac{x}{3\pi}\right)\left(1 - \frac{x}{4\pi}\right) \cdots$$

$$= \left(1 - \frac{x^2}{\pi^2}\right)\left(1 - \frac{x^2}{2^2\pi^2}\right)\left(1 - \frac{x^2}{3^2\pi^2}\right)\left(1 - \frac{x^2}{4^2\pi^2}\right) \cdots$$

$\sin x / x$ is, in a sense, an "endless polynomial" in powers of $x^2$ rather than just $x$, and the coefficient of $x^2$ is $-1/3!$. Hence, if everything we have done so far turns out to be correct, then we would expect to find that

$$-\frac{1}{3!} = -\left(\frac{1}{\pi^2} + \frac{1}{2^2\pi^2} + \frac{1}{3^2\pi^2} + \frac{1}{4^2\pi^2} + \frac{1}{5^2\pi^2} + \cdots\right)$$

that is

$$\frac{\pi^2}{6} = \sum_{r=1}^{\infty} \frac{1}{r^2}$$

We have made a lot of dubious moves, but if this value agrees with the available evidence (including our calculation of partial sums $\sum_{r=1}^{N} (1/r^2)$), then it is hard to believe that it could actually be *wrong*[7]! Why the procedure has worked in this case remains a serious question—especially since we would obviously like to use it to calculate other infinite sums.

Footnotes to Exercise 2

(1) The belief that any polynomial of degree $n$ has $n$ roots originated in the early seventeenth century, but it was first proved by Gauss (1777–1855) in 1797[5]—fourteen years after Euler's death.

(2) There is no obvious reason why we should not write down an "endless polynomial" if we so wish. But it is not at all clear what it means, whether, or how, such endless polynomials can be safely manipulated, and so on.

(3) Again, we can write down an "endless product" any time we like. But what does it mean? How do I multiply out the brackets? (What is the "constant term" in

$$(x - 1)(x - \tfrac{1}{2})(x - \tfrac{1}{4})(x - \tfrac{1}{8}) \cdots ?$$

What is the coefficient of $x$?)

(4) Starting from

$$p(x) = (x - \alpha_1)(x - \alpha_2) \cdots (x - \alpha_n)$$

we can divide through by $(-\alpha_1) \cdot (-\alpha_2) \cdot \cdots \cdot (-\alpha_n) = a_0$ to get

$$p(x) = a_0\left(1 - \frac{x}{\alpha_1}\right)\left(1 - \frac{x}{\alpha_2}\right) \cdots \left(1 - \frac{x}{\alpha_n}\right).$$

(5) This idea is extremely optimistic. Many "endless polynomials" such as

$$1 + x + x^2 + x^3 + x^4 + \cdots = \frac{1}{1 - x}$$

or

$$1 + x + \frac{x^2}{2!} + \frac{x^3}{3!} + \frac{x^4}{4!} + \cdots = e^x$$

[5]The reader is encouraged to question all such authoritarian assertions: Gauss's first two "proofs" are outlined and criticised in, for example, Felix Klein's book *Development of Mathematics in the Nineteenth Century* (first published by Springer in 1928, and recently published in an English translation by Math Sci Press, 1979).

have no "roots" at all, and so can certainly not be factorised in this way! Other "endless polynomials," such as

$$1 + \sqrt{2}x + \frac{\sqrt{2}(\sqrt{2}-1)}{2!} x^2 + \cdots = (1+x)^{\sqrt{2}}$$

may have "roots" (in this case $x = -1$ is the only root), but can certainly not be written as a product of linear factors.

(6) Once we recognise this "endless polynomial" as our old friend $sin$ then it is obvious that the only *real* "roots" are

$$0, \quad \pm\pi, \quad \pm 2\pi, \quad \pm 3\pi, \quad \text{etc.}$$

But might there not be other *complex* "roots"? (If not, why not?)

(7) Work out $\pi^2/6$ and compare it with your available evidence!

3. Use the fact that

$$\sum_{r=1}^{\infty} \frac{1}{r^2} = \frac{\pi^2}{6}$$

to show that the probability that two positive whole numbers $m$ and $n$ chosen at random will have no common factors is $6/\pi^2$. [Hint: What is the probability that $m$ and $n$ are *not both* divisible by the prime $p$?]

# PART II
# NUMBER

In which
- we examine the ordinary *finite* processes involved in geometric measurement, and in the representation of numbers as ordinary decimals and as continued fractions;
- we discover line segments which cannot be measured and numbers which cannot be represented as decimals or as continued fractions unless we allow *infinite* processes, *endless* decimals and *endless* continued fractions;
- and we investigate how these *infinite* processes, *endless* decimals and *endless* continued fractions can be given a clear meaning, and how they can be safely manipulated.

# Mathematics: Rational or Irrational?

> The real importance of the Greeks for the progress of the
> world is that they discovered the almost incredible secret that
> the speculative Reason was itself subject to orderly methods.
>
> A.N. Whitehead, *The Function of Reason*

Perhaps the most basic idea in all of mathematics is that of *counting numbers*—the positive whole numbers. If human beings are to get interested in anything mathematical, then we should not be surprised to find them beginning with these counting numbers—their patterns of odd and even; the squares, cubes, and higher powers; the triangular numbers 1, 3, 6, 10, 15, ...; the primes; the divisors of a given number and its factorisation as a product of primes; and many other fascinating properties (see, for example, Exercise 2).

Such was the direction apparently taken by the Pythagoreans. They were a curious collection of men, who formed a mathematico-religious community around 530 BC in *Croton*—a seaport in Southern Italy, which was at the time part of Magna Graecia. We possess no actual record of what they did, and many of the discoveries attributed to them by later Greek writers are nowadays believed to be somewhat misleading. But it is quite clear that their ideas had a significant influence on the direction taken by later Greek mathematics. What follows does not pretend to be an accurate historical example: we simply exploit the Pythagoreans as a means of helping the reader to reflect on the enormity of the gulf between

*counting numbers and fractions—that is, the positive rational numbers,*

and

*those other numbers like $\sqrt{2}$, $\pi$, $e$,*

which more often than not we simply call *irrational*—and then ignore!

The Pythagoreans' study of the properties of ordinary counting numbers was neither a cold, dispassionate pursuit, nor a mere hobby: they were committed to explaining the whole universe in terms of such numbers.[1] In

---

[1] This may seem a trifle optimistic to us, and certainly *the way* they hoped to do this was too simple-minded. But the idea is remarkably similar *in spirit* to the successful attempt (around 1870 AD) to develop the whole of elementary calculus from the arithmetic of ordinary whole numbers—the so-called *arithmetisation of analysis*.

this they were encouraged by a number of really very impressive discoveries—the best known of which is perhaps that *the musical harmonies are expressed by ratios of whole numbers*. For example, if the length of a taut string is halved by fixing its midpoint in some way, then the note it emits on being plucked goes up by precisely one octave. Thus a pair of notes *in unison*, one whole octave apart, corresponds in some very real sense to the ratio 2 : 1, and this is completely independent of the original length of the string. In the same way if we reduce the length of a taut string by one third, the note it emits on being plucked goes up by a *fifth* (say from C to G); thus a pair of notes separated by an interval of a *fifth* corresponds to the ratio 3 : 2. Similarly an interval of a *fourth* (say from C to F) corresponds to the ratio 4 : 3, and so on. Since it was precisely such intervals as these which sounded *harmonious* to the Pythagorean ear, they naturally concluded that there was more to whole numbers than met the eye.

This commitment to whole numbers and their ratios did not prevent them from making considerable progress in geometry. If one sticks to whole numbers and insists on a *fixed* unit, then one can only really compare segments which are whole multiples of the fixed unit—a restriction which seems unlikely to lead to much interesting geometry. But with a little more imagination one may realise that a *fixed* unit is an unnecessary luxury: one can in practice compare two segments, AB, CD in terms of whole numbers whenever there exists a third segment—say MN—*in terms of which both AB and CD can be measured exactly*. Thus in Figure 2 we see that the segments

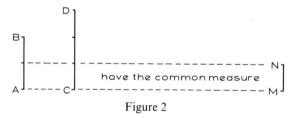

Figure 2

AB and CD are in the ratio 2 : 3 in the sense that we can find a segment MN which fits AB exactly twice and CD exactly 3 times. The important change is that we no longer insist on sticking to one and the same unit MN for all time: instead, given any two segments AB, CD, *we are free to choose any segment MN* which is sufficiently small to suit both AB and CD.

One of the many results the Pythagoreans discovered in their study of geometry is what we now call Pythagoras' theorem.[2] This result is remarkable because it is not one of those facts that can be stumbled on by accident; nor is it as intuitively obvious as, for example, the fact that the

---

[2] The Babylonians were aware of this result more than a thousand years before Pythagoras. In practice, many results and ideas which are traditionally attributed to a particular mathematician were not in fact originally due to the person named (e.g. *Thales'* theorem, *Taylor* series, *Euler's* formula, *de Moivre's* theorem, *Napier*ian logarithms, the *Argand* diagram).

sum of any two sides of a triangle is always greater than the third side. Thus one must assume that the Pythagoreans had some kind of *proof*, or *plausible explanation*, as to why the result is true, though we do not know what it was. Nevertheless many historians have suggested possible *"Pythagorean proofs"* of Pythagoras' theorem: for example, consider Figure 3. *In what*

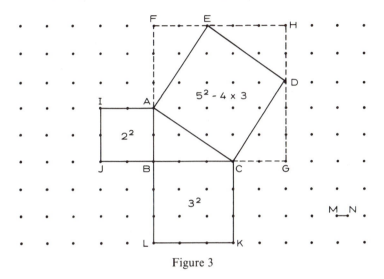

Figure 3

*sense* might this diagram correspond to a *proof* in the eyes of a highly critical Pythagorean? *You* may feel that because *specific* line segments are shown, the diagram can only apply to a *specific* case—namely, a right angled triangle with sides of length 2, 3 and $\sqrt{13}$—and so could not possibly represent a *general* proof. But is this entirely fair? The diagram not only constructs the square ACDE for this case, but suggests clearly how to construct a square on any line segment AC joining two dots in the square dot-lattice: if the route from A to C takes us

m steps down and n to the right

then from C to D we go

m steps to the right and n steps up

and from D to E we go

m steps up and n to the left.

Moreover, this procedure not only constructs the square ACDE, but also produces four *visibly congruent* right-angled triangles ABC, CGD, DHE, EFA, and the giant $m + n$ by $m + n$ square BGHF. Thus, although *the detailed shape* of the diagram depends on the fact that the ratio of AB to BC is 2 : 3, *the method it suggests* will work equally well for any other ratio

$m : n$. Thus whatever triangle ABC we start with, we will always have

$$area \text{ ACDE} = area \text{ FBGH} - 4 \times (area \text{ ABC})$$

which we simply translate in each case as follows:

|  |  |
|---|---|
| *Particular case* | *General case* |

$$
\begin{aligned}
\text{AC}^2 &= (3 + 2)^2 - 4 \times (\tfrac{1}{2} \times 2 \times 3) \\
&= 3^2 + 2^2 \\
&= \text{BC}^2 + \text{AB}^2
\end{aligned}
\qquad
\begin{aligned}
\text{AC}^2 &= (\text{BC} + \text{CG})^2 - 4 \times (\tfrac{1}{2} \times \text{AB} \times \text{BC}) \\
&= (\text{BC} + \text{AB})^2 - 2 \times \text{AB} \times \text{BC} \\
&= \text{BC}^2 + \text{AB}^2
\end{aligned}
$$

Moreover, the areas ABJI, BCKL, FBGH, ACDE which are being compared *all have a common measure*, since they are all whole number multiples of one of the basic unit squares: thus the areas of these four squares are in the ratio

|  |  |
|---|---|
| *Particular case* | *General case* |
| $4 : 9 : 25 : 13$ | $m^2 : n^2 : (m + n)^2 : m^2 + n^2$ . |

So far then, so good! But if you could really bring yourself to believe, like the Pythagoreans, that *whole numbers and ratios of whole numbers* hold the key to the whole physical and spiritual universe, then you too would experience the same profound shock at what follows. So try to imagine just for the moment

(1) that *number* means positive whole number;
(2) that geometry is concerned with *relationships between* geometrical line segments, or other figures—and *not* with numerical lengths and areas;
(3) that our understanding of geometry depends on the assumption that *any two line segments, or shapes, being compared can be measured with a common measure*;

and last, but not least

(4) that your whole life is moulded by a religion in which each whole number has its own social or religious meaning (e.g. 1 represents the divine, 4 is the number of justice, 5 is the number of marriage, and so on).

Let us then return to Figure 3 and try to write out a *general* proof of Pythagoras' theorem on the lines suggested above. Such a proof might run as follows:

Given any right-angled triangle ABC with hypotenuse AC.
*The line segments AB and BC must have a (suitably small) common measure*–MN say (Figure 4).
We may thus construct the square dot-lattice using *MN* as the basic unit.

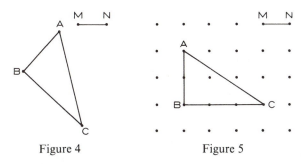

Figure 4                    Figure 5

We can then draw a copy of the triangle *ABC* on this lattice with its vertices at lattice points (Figure 5).
The proof represented by Figure 3 then applies to our triangle ABC.

Up to this point there is no indication that we have done anything at all unreasonable, and the indication that is about to appear is not exactly direct! To simplify the discussion, let us apply our general proof to the simplest possible right-angled triangle—namely, half a square, in which AB = BC (Figure 6).

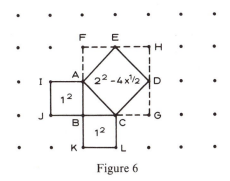

Figure 6

Given a right-angled triangle ABC with AB = BC.
AB and BC have the common measure MN = AB.
We may construct the dot lattice with AB as basic unit and draw a copy of ABC on this lattice (Figure 6).
The proof represented by Figure 3 now shows that $AC^2 = BC^2 + AB^2$.

There is still no visible sign that anything has gone wrong. But what if some curious Pythagorean realised that the assumption that *the line segments AB and BC must have a (suitably small) common measure* should also be true of the line segments AB and AC in Figure 6? After all, since the line segments AB and AC clearly exist, we can easily construct a right-angled triangle ABC' with hypotenuse BC' and sides AB and AC' = AC (Figure 7).

If we now wish to apply our general argument to the triangle ABC' in

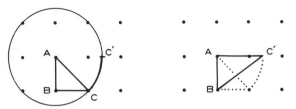

Figure 7

Figure 7, then we should have to assume that the sides AB and AC' have some (suitably small) common measure—say M'N'. Thus M'N' must go exactly into AB (say $m$ times), and it must also go exactly into AC' (say $n$ times). But this simply means that the ratio of AB to AC' is precisely $m : n$.

$$AB : AC' = m : n$$

so        $$AB^2 : AC'^2 = m^2 : n^2.$$

But $AC'^2 = AC^2$, and the squares ABIJ on AB, and ACDE on AC can both be measured with one of the basic unit squares (see Figure 6): their areas $AB^2$ and $AC^2$ are in the ratio 1 : 2. Thus

$$AB^2 : AC^2 = 1 : 2,$$

so        $$m^2 : n^2 = 1 : 2,$$

so        $$2m^2 = n^2.$$

But if $n^2 = 2m^2$, then $n$ cannot be odd—otherwise $n^2$ would have to be odd; thus $n$ is *even*, so $n$ is twice some whole number—say $n = 2n_1$. But then

$$2m^2 = (2n_1)^2$$

so        $$m^2 = 2 \cdot n_1^2 .$$

But if $m^2 = 2n_1^2$, then $m$ also cannot be odd—otherwise $m^2$ would have to be odd; thus $m$ is *even*, so $m$ is twice some whole number—say $m = 2m_1$. But then

$$(2m_1)^2 = 2 \cdot n_1^2$$

so        $$2 \cdot m_1^2 = n_1^2 .$$

But now $n_1$ cannot be odd, so $n_1 = 2 \cdot n_2$ for some whole number $n_2$, and

$$2 \cdot m_1^2 = (2n_2)^2$$

so        $$m_1^2 = 2 \cdot n_2^2 .$$

But now $m_1$ cannot be odd, so $m_1 = 2 \cdot m_2$ for some whole number $m_2$, and

$$(2m_2)^2 = 2 \cdot n_2^2$$

$$\text{so} \qquad 2 \cdot m_2^2 = n_2^2$$

$$\vdots$$

This procedure could clearly go on for ever and ever, starting with the ordinary whole numbers $m$, $n$ and proceeding

from the *whole* numbers $m$, $n$          to the *whole* numbers
                                            $m_1 = m/2$, $n_1 = n/2$
from the *whole* numbers $m_1$, $n_1$       to the *whole* numbers
                                            $m_2 = m/2^2$, $n_2 = n/2^2$
from the *whole* numbers $m_2$, $n_2$       to the *whole* numbers
                                            $m_3 = m/2^3$, $n_3 = n/2^3$
from the *whole* numbers $m_3$, $n_3$       to ...

and so on ad infinitum.

At this point (if not before) our imaginary Pythagorean may begin to feel decidedly uneasy. Something has certainly gone very wrong indeed; for we seem to have produced an endless decreasing sequence

$$m, \quad m_1 = \frac{m}{2}, \quad m_2 = \frac{m}{2^2}, \quad m_3 = \frac{m}{2^3}, \quad m_4 = \frac{m}{2^4}, \quad \dots \text{ ad. inf.}$$

of *positive whole* numbers, and this is clearly impossible! However large the whole number $m$ may be, we will inevitably reach a point where $2^k > m$; but then $m_k = m/2^k$ is less than 1, so cannot possibly be a whole number. We are confronted with a *paradox*.

Now it would be very convenient if we were able to say simply that a *paradox* is an argument or a conclusion which only appears absurd, but which can actually be explained fairly easily. In most instances this is certainly true. But mathematics repeatedly witnesses the discovery of more serious paradoxes whose resolution necessitates the *reconstruction* of some considerable portion of contemporary mathematics. One of the ways in which mathematics has to grow is by struggling to resolve such paradoxes or contradictions.

Given the benefit of hindsight we can present this Pythagorean paradox in a way which no longer seems at all paradoxical:

(i)  we began by assuming that whole numbers and their ratios were sufficient to describe the relationship between any two line segments AB, CD;
(ii) we then used this assumption to "prove" a beautiful theorem about right-angled triangles (Figure 3);
(iii) applying this result to the simplest right-angled triangle of all (ABC in Figure 6) we constructed another right-angled triangle ABC' (Figure 7);
(iv) interpreting our original assumption (i) for the particular triangle ABC',

we produced the endless sequence $m, m_1, m_2, m_3, \ldots$ of whole numbers;

(v) this evidently false conclusion means simply that our original assumption (i) is in fact untenable (and, in particular, that we have not yet proved Pythagoras' theorem).

But ideas which we have come to take for granted are not easily abandoned.

EXERCISES

1. Obtain, or construct for yourself, two A4 sheets of $\frac{1}{2}$cm $\times$ $\frac{1}{2}$cm square dot-lattice. In this chapter we have seen how to construct squares *with corners at lattice points*, and having areas 1, 2, 4, 9, 13, 25 units (the unit here will be a $\frac{1}{2}$cm $\times$ $\frac{1}{2}$cm basic square)

(i) For which whole numbers $n \leq 60$ can one construct such a square, with corners at lattice points, and having area $n$ units?

(ii) What is the smallest whole number $n$ which arises thus in two (or more) entirely different ways?

(iii) Which numbers arise thus in precisely one way?

2. (i) In the following table the first few primes have been underlined; find all the other primes and underline them (there are *thirty* altogether):

```
  0   1   2   3   4   5   6   7   8   9  10  11  12  13  14  15  16  17  18  19
 20  21  22  23  24  25  26  27  28  29  30  31  32  33  34  35  36  37  38  39
 40  41  42  43  44  45  46  47  48  49  50  51  52  53  54  55  56  57  58  59
 60  61  62  63  64  65  66  67  68  69  70  71  72  73  74  75  76  77  78  79
 80  81  82  83  84  85  86  87  88  89  90  91  92  93  94  95  96  97  98  99
100 101 102 103 104 105 106 107 108 109 110 111 112 113 114 115 116 117 118 119
```

(ii) Circle each prime which can be written as the sum of *two* squares (for example: $2 = 1^2 + 1^2$, $5 = 2^2 + 1^2$ etc.).

(iii) Make an intelligent guess as to which primes can, and which primes cannot, be written as a sum of two squares.

(iv) Test your guess as follows. Consider the next few lines of the above table. The first two primes have been underlined; find all the other primes and underline them (there are *eleven* altogether):

```
120 121 122 123 124 125 126 127 128 129 130 131 132 133 134 135 136 137 138 139
140 141 142 143 144 145 146 147 148 149 150 151 152 153 154 155 156 157 158 159
160 161 162 163 164 165 166 167 168 169 170 171 172 173 174 175 176 177 178 179
```

Now *use your guess* to predict which of these eleven primes can, and which cannot, be written as the sum of two squares. Then check whether your predictions are correct. Revise your guess if necessary.

3. We derived a contradiction by assuming that the segments AB and AC in Figure 6 have a common measure. Write out the corresponding proof that the segments AB and AC in Figure 3 have no common measure. (The proof starts out exactly as for Figure 6:

"Suppose AB and AC have a common measure: call it MN.

If MN goes exactly $m$ times into AB and exactly $n$ times into AC, then the ratio of AB to AC is just $m : n$;

$$AB : AC = m : n$$

so     $$AB^2 : AC^2 = m^2 : n^2 .$$

But we saw that both the square ABIJ on AB and the square ACDE on AC can be measured with one of the basic unit squares and that their areas $AB^2$ and $AC^2$ are in the ratio $4 : 13$

$$AB^2 : AC^2 = 4 : 13$$

so   $$m^2 : n^2 = 4 : 13 \quad \text{...}"$$

In the proof for Figure 6 we obtained

$$m^2 : n^2 = 1 : 2$$

and then changed this to

$$2m^2 = n^2$$

before arguing in terms of *odd* and *even* numbers.
In the proof for Figure 3 we obtain

$$13 \cdot m^2 = 4 \cdot n^2$$

and must argue in terms of *multiples of 13* rather than multiples of 2.
If $4n^2$ is a multiple of 13, what does this tell us about $n$, and why?)

4. Let ABC be a right-angled triangle with hypotenuse AC with its vertices A,B,C at lattice points in the square dot-lattice.

(i) If AB = 1 unit and BC = 2 units, do AB and AC have a common measure?

(ii) If AB = 1 unit and BC = $m$ units, for which values of $m$ do AB and AC have a common measure?

(iii) If AB = 2 units and BC = $m$ units, for which values of $m$ do AB and AC have a common measure?

(iv) If AB = 3 units and BC = $m$ units, for which values of $m$ do AB and AC have a common measure?

5. For which triples $l$, $m$, $n$ of whole numbers is it possible to construct a right-angled triangle ABC, with

$$AB = l \text{ units}, \quad BC = m \text{ units}, \quad AC = n \text{ units}?$$

[Hint: You are given two pieces of information: (i) $l, m, n$ are whole numbers; (ii) $l^2 + m^2 = n^2$. The first piece of information suggests that you must use facts about the factorisation and divisibility of whole numbers. The second piece of information gives you something to use these on.

*Step 1*: If we write $hcf(l, m)$ for the highest common factor of $l$ and $m$, show that

$$hcf(l, m) = hcf(m, n) = hcf(n, l)$$

*Step 2*: If $hcf(l, m) = h$, then $l = hl'$, $m = hm'$, $n = hn'$, and (i) $l'$, $m'$, $n'$ are whole numbers, (ii) $l'^2 + m'^2 = n'^2$, and (iii) $hcf(l', m') = hcf(m', n') = hcf(n', l') = 1$.

*Step 3*: Show that $n'$ is odd, whence one of $l'$, $m'$ is odd ($m'$ say) and the other is even ($l'$ say).

*Step 4*: $l'^2 = (n' - m')(n' + m')$ and $hcf(n' - m', n' + m') = 2$.

*Step 5*: $(n' - m')/2 = q^2$ and $(n' + m')/2 = p^2$ are both squares, whence $n' = p^2 + q^2$, $m' = p^2 - q^2$, $l' = 2pq$, where $p$ and $q$ are relatively prime integers of opposite parity.]

CHAPTER II.2

# Constructive and Non-constructive
# Methods in Mathematics

In the previous chapter we started by tacitly assuming that *any* given pair of line segments could necessarily be measured *exactly* by some (suitably small) *common measure*. But, when this assumption was applied to the pair of segments AB, AC in Figure 6, we derived a contradiction. We were thus forced by the principles of logic to admit that our tacit assumption was untenable: there clearly exist *some* pairs of segments (such as AB, AC in Figure 6) which have *no common measure at all*.

This remarkable fact influenced not only the theoretical approach of the ancient Greeks, but is to a large extent responsible for the two hundred year gap between the invention of the methods of the calculus around 1670 and their eventual clarification around 1870. In modern language this simple *fact* can be interpreted as saying that the *arithmetic* of whole numbers and their ratios is inadequate as a *numerical* description of what goes on in *geometry*. Geometry certainly *includes* the arithmetic of (positive) whole numbers and their ratios, since for any pair $m$, $n$ of (positive) *whole* numbers one can find segments AB, CD for which

$$AB : CD = m : n,$$

and the addition and multiplication of such ratios have geometrical interpretations. But geometry seems to be much more than mere arithmetic, since it also gives rise to pairs of segments which have no simple numerical description in terms of whole numbers or their ratios. Thus to the Greeks, *geometry* (the study of points and straight lines) appeared to be *far more general* than *arithmetic*; they therefore chose geometry as the natural framework within which to develop their mathematics. But they still had to find some way of accommodating those awkward pairs of segments like AB, AC in Figure 6, which have no common measure, and which therefore cannot be compared in terms of ratios of whole numbers. The remarkably subtle geometrical method developed by Eudoxus (around 350 B.C.) for handling such pairs of segments with no common measure, is recognised as one of the greatest achievements of classical Greek mathematics.

Now the calculus is concerned with

*numbers* and *functions*

in a way which cannot be easily reduced to the geometry of points and straight lines. But for a long time the *arithmetical* interpretation of the calculus was overlooked—partly because of the lasting impact of the Greek geometrical approach,[1] and partly because the origins and applications of the calculus in the seventeenth and eighteenth centuries have a strong geometrical flavour.[2] Around 1860 Richard Dedekind (1831–1916) realised that Eudoxus' geometrical method for handling pairs of segments with no common measure could be reinterpreted *in a purely arithmetical way* as a way of constructing *the real numbers*—rationals *and* irrationals—by starting from the rational numbers alone.

But, for all its *theoretical* significance, the *necessary existence* of pairs of line segments having no common measure, which was proved in the previous chapter, is no help at all in deciding whether or not two particular segments AB, CD have a common measure; and if they have, then it suggests no method by which we might actually find such a common measure. A somewhat analogous situation arises in the study of prime numbers. It is fairly easy to prove (Exercise 1) that there must exist infinitely many prime numbers. This remarkable fact affects our whole theoretical approach to the study of whole numbers. But it is no help at all in deciding whether or not a given number (like 65,537) is prime; nor does it help us to find the 1001st prime number.

Now these two examples—common measures for pairs of line segments, and the search for prime numbers—demonstrate the distinction between two important strategies in mathematics: the *existential* approach and the *constructive* approach. The existential approach is based on cold logic and often uses *indirect* arguments—that is, in order to prove that some result is true we suppose instead that it is false, and then derive a contradiction (from which we conclude that the result cannot possibly be false, and so must be true after all). Though such proofs can be very elegant, a certain psychological maturity seems to be required before they become entirely convincing. It is perhaps for this reason that very little of the mathematics one meets in school is of this kind—and that which is of this kind often goes down rather badly. School mathematics is *naively constructive* in spirit: no one bothers to discuss whether or not the sum of two whole numbers exists, one simply goes ahead and adds them; we spend relatively little time discussing the existence or non-existence of a point of intersection of two

---

[1] The classical Greek works became generally available in Europe only during the course of the sixteenth century and thereafter. The impact of these rediscovered works, and the reverence in which they were held, should not be underestimated.

[2] For example, the *geometry of curves and surfaces* (such as the construction of tangents and normals, the computation of areas, volumes, surface areas, centres of gravity, etc.), and the *geometry of physical systems* (such as the shape of a chain hanging under its own weight, the shape of the water's surface in a rotating bucket, the motion of planets in the solar system, etc.).

straight lines in the plane, we simply treat the equations of the two lines as simultaneous equations, and solve them if we can—if we fail, then the lines must be parallel; we do not concern ourselves too much as to whether the roots of a given polynomial really exist, but simply calculate them approximately using some iterative method like Newton-Raphson. A *genuinely constructive* approach differs from the approach implicit in these examples by being more interested in the effectiveness and rigorous justification of the procedures used, and in their theoretical consequences; it differs from the existential approach in not being content to work with ideas whose only justification arises from a mere existence proof, but insists that *the objects corresponding to these ideas be constructible* in some mathematical sense.

### EXERCISES

1. (i) Let $p$ be a prime number and $N$ any whole number. Explain why $p$ cannot exactly divide the number

$$p \cdot N + 1 \, .$$

(ii) Let $p_1, p_2, \ldots, p_n$ be prime numbers. Explain why none of the prime factors of the number

$$p_1 \cdot p_2 \cdot \ldots \cdot p_n + 1$$

can occur in the list $p_1, p_2, \ldots, p_n$.

(iii) Write out a proper proof (by induction) of the following statement:

$P(n)$: *"If $p_1, p_2, \ldots, p_n$ are prime numbers, then there always exists a prime number $p_{n+1}$ different from all of $p_1, p_2, \ldots, p_n$."*

(iv) Let $p_1 = 2$.
  Let $p_2$ be the smallest prime factor of $p_1 + 1$. Find $p_2$.
  Let $p_3$ be the smallest prime factor of $(p_1 \cdot p_2) + 1$. Find $p_3$.
If $p_1, p_2, \ldots, p_n$ have been defined, let $p_{n+1}$ be the smallest prime factor of $(p_1 \cdot p_2 \cdot \ldots \cdot p_n) + 1$.
Find $(p_1, p_2, p_3,) p_4, \ldots, p_{10}$.

# Common Measures, Highest Common Factors and the Game of Euclid

> ... the whole structure has just one cornerstone,
> namely the algorithm by which one calculates
> the highest common factor of two whole numbers.
>
> P.G. Lejeune Dirichlet, *Number Theory*

In Chapter II.1 we gave an *indirect* proof of the non-*existence* of a common measure for certain pairs of line segments (such as the side and diagonal of a square). In this chapter we shall develop a *direct, constructive* procedure for finding a common measure of two segments when a common measure exists. In the next chapter we shall complement the discussion of Chapter II.1 by using this *constructive* procedure to give a second proof that the side and diagonal of a square do not have a common measure.

Our constructive procedure is based on the Euclidean method for finding the highest common factor of two whole numbers (Euclid Book VII, Proposition 2). In Euclid Book X (Propositions 2, 3) this method is interpreted as providing criteria for deciding whether a pair of line segments does or does not admit a common measure, and this is done in much the same way as at the end of *this* chapter. However it is not at all clear that the application of these ideas which we shall discuss in the *next* chapter would have been familiar to the ancient Greeks, though there are grounds for supposing that it might have been.

Suppose we are presented with two line segments AB and CD, which have a common measure MN: that is, there exist whole numbers $a$ and $b$ such that

$$AB : MN = a : 1 \quad \text{and} \quad CD : MN = b : 1 .$$

Then although MN is *a* common measure of AB and CD, it may not be *the greatest* common measure: for example, if

$$AB : MN = 50 : 1 \quad \text{and} \quad CD : MN = 45 : 1 ,$$

we could use a segment 5 "MN units" in length to measure both AB and CD. But though it is not immediately obvious, it is not hard to show that no matter which two line segments AB, CD we start with, if they have a common measure MN, then the length in "MN units" of the greatest common measure of our two segments AB, CD is given by the *highest common factor* of the two whole numbers *a, b*—which we denote simply by *hcf*(*a, b*) (see Theorem 2.3 below). Thus any effective method of finding the highest common factor *hcf*(*a, b*) of two whole numbers *a, b* should yield a *direct, constructive* procedure for actually finding the greatest common measure of two line segments—when, that is, they actually have a common measure.

By way of an introduction to Euclid's delightfully simple method for finding *hcf*(*a, b*) for any two whole numbers *a, b*, we shall digress briefly to introduce "the game of Euclid." This is a game for two players called A and B, which *really needs to be played rather than imagined*—so find yourself an opponent, and play until you have a reasonable feel for which moves pay and which do not.

| **Rules** | **Specimen game** | |
|---|---|---|
| (1) Each player thinks of a number and records it secretly. | 27 (A) | 289 (B) |
| (2) Players toss to decide who starts. | Player A to start | |
| (3) Both players show their numbers *a,b*, which are then written as an ordered pair (*a,b*). | (27, ↓A (27, ↓B | 289) 46) |
| (4) Players take turns to move. The first player can transform the pair (*a,b*) by subtracting any multiple of the smaller number from the larger. The second player does the same with the new pair. | (27, ↓A (8, ↓B (8, ↓A (2, | 19) 19) 3) 3) |
| (5) The first player to produce a pair involving 0 wins. | ↓B (2, ↓A (0, | 1) 1)   A wins |

This is clearly not just a game of chance. There is in fact a simple way of deciding at the very start which of the two players can definitely win assuming that he makes no false move. But though the secret is simple, it is not at all obvious, and you should use the Exercises to analyse the game more fully. Meantime we shall restrict ourselves to a few elementary questions:

What would happen if the two players tossed to see who should start *before* writing down their chosen numbers? (Suppose I know I am to start, which number should I choose?)

Suppose I *always* choose the number 1, how often will I win? What if I always choose the number 2? (Some such practices may spoil the game. Do you think an extra rule is necessary? If so, can you suggest one?)

If one of the initial numbers $a$, $b$ is a 1, then the first player clearly wins. What is the simplest initial number pair for which the first player always loses?

In the *Specimen game* above, B has no choice in his first and third moves. However B's second move could have been $(8, 19) \rightarrow (8, 11)$. Would this have been a better choice?

In the *Specimen game* above both A and B are guilty of unforced errors. Suppose the game is replayed in the light of experience—starting from $(27, 289)$ with A to move first. Write out the sequence of moves the players *should* make.

*Finally, and most importantly for this section, what is the significance of the non-zero number in the winning pair?*

Why, you may ask, is this final question so important? Well, though it may not help the player who is interested only in "the game of Euclid," we shall see that the non-zero number in the winning pair is in fact the *highest common factor* of the two numbers we started with: thus the winning pair $(0, 1)$ in our *Specimen game* reflects the fact that the highest common factor of $27 \, (= 3^3)$ and $289 \, (= 17^2)$ is just 1.

"The game of Euclid" is in fact a restatement of the Euclidean method for finding the highest common factor of two whole numbers, but in the game of Euclid the mathematical focus is neatly switched away from the *end result*, and onto the *process* by which that result is obtained. However *we* must now turn our attention back to the end result: the highest common factor $hcf(a, b)$ of two whole numbers $a$, $b$.

The Euclidean method for finding the highest common factor of two whole numbers is based on two very simple observations:

(i)   *if $h$ is a factor of $a$*, (that is, $a = h \cdot m$ for some whole number $m$),
      and $h$ is a factor of $b$, (that is, $b = h \cdot n$ for some whole number $n$),
      *then $h$ is also a factor of $a - b$*, $(a - b = hm - hn = h \cdot (m - n))$.
(ii)  *if $h'$ is a factor of $b$*, (that is, $b = h' \cdot n'$ for some whole number $n'$),
      and $h'$ is a factor of $a - b$, (that is, $a - b = h' \cdot m'$ for some whole number $m'$),
      *then $h'$ is also a factor of $a$*, $(a = (a - b) + b = h' \cdot m' + h' \cdot n' = h'(m' + n'))$.

From these it follows that the *common* factors of $a$ and $b$ are precisely the same as the *common* factors of $a - b$ and $b$, whence

$$hcf(a, b) = hcf(a - b, b). \qquad (2.3.1)$$

In particular since each move in the game of Euclid transforms the given pair in precisely this way, the highest common factor of the *initial* pair is the same as the highest common factor of *each subsequent* pair; since the final (winning) pair is always of the form $(c, 0)$ or $(0, c)$, to prove our assertion about the non-zero number in the winning pair we need only check that

$$hcf(c, 0) = c.$$

Now clearly $c$ is the largest factor of $c$; *but is $c$ really a factor of* $0$? Fortunately the answer is "*Yes*"! [If this fact comes as something of a surprise, consider carefully what it means to say that the whole number $c$ is a factor of the whole number $n$:

$$c \text{ is a factor of } n$$

means precisely that we can factorise $n$ as a product

$$n = c \times m$$

for some whole number $m$; but if $n = 0$, then whatever the value of $c$, we can always factorise 0 as a product

$$0 = c \times 0;$$

in other words

$$c \text{ is a factor of } 0.]$$

If this does not strike you as being a very efficient way of working out highest common factors, try to find some *other* way of computing $hcf(5776, 9633)$.

$$* \quad * \quad * \quad * \quad *$$

Now consider the following solution based entirely on repeated use of the relation (2.3.1):

$$a = 5776, \qquad b = 9633$$

(1) $hcf(5776, 9633) = hcf(5776, 9633\text{-}5766) = hcf(5776, 3857)$
(2) $hcf(5776, 3857) = hcf(5776 - 3857, 3857) = hcf(1919, 3857)$
(3A) $hcf(1919, 3857) = hcf(1919, 3857 - 1919) = hcf(1919, 1938)$
(3B) $hcf(1919, 1938) = hcf(1919, 1938 - 1919) = hcf(1919, 19)$

Steps (3A) and (3B) can clearly be combined by writing
(3) $hcf(1919, 3857) = hcf(1919, 3857 - 2 \times 1919) = hcf(1919, 19)$

Similarly
(4) $hcf(1919, 19) = hcf(1919 - 101 \times 19, 19) = hcf(0, 19) = 19.$

The advantage of this method lies not in its brevity (consider, for example, $hcf(2584, 1597)$), but rather in its repetitive simplicity and its complete generality: to appreciate this you should work through the examples of

Exercise 4, and if possible express the method in the form of a simple program for a programmable calculator so that you can test the simplicity and effectiveness of the method even for pairs of very large numbers.

Finally, in this section, we apply these ideas to obtain a constructive procedure for solving our original problem:

> If two given line segments AB, CD
> have a common measure, how can we
> actually find such a common measure?

The procedure we use is precisely that used in the game of Euclid, and in the Euclidean method for finding the highest common factor of two whole numbers $a$, $b$; this produces not only *a* common measure but *the greatest* common measure of AB and CD.

Imagine the two segments AB, CD arranged as in Figure 8. The pro-

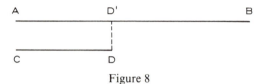

Figure 8

cedure is based, just like the Euclidean method for *hcf*'s, on two very simple facts about the segments AB, CD and D'B in Figure 8:

(i) *if* MN measures AB exactly (that is, $AB : MN = m : 1$ for some whole number $m$),
    and MN measures CD exactly (that is, $CD : MN = n : 1$ for some whole number $n$),
    *then* MN measures $D'B = AB - CD$ exactly;

(ii) *if* M'N' measures CD exactly (that is, $CD : M'N' = m' : 1$ for some whole number $m'$),
    and M'N' measures $D'B = AB - CD$ exactly (that is, $D'B : M'N' = n' : 1$ for some whole number $n'$),
    *then* M'N' measures AB exactly.

From these it follows that

> the common measures of AB and CD

are precisely the same as

> the common measures of $D'B = AB - CD$ and CD;

in particular (see Exercise 7)

> the *greatest* common measure of AB and CD

is precisely the same as

> the *greatest* common measure of D'B and CD.

We can now start with the pair of segments D'B, CD and imitate the previous step

from the pair AB, CD        to the pair D'B = AB − CD, CD,

by passing now

from the pair D'B, CD        to the pair D"B = D'B − CD, CD

as in Figure 9.

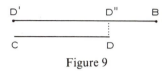

Figure 9

As before,

the common measures of D'B and CD

are precisely the same as

the common measures of D"B and CD,

so we can repeat the process once more—starting this time with the pair of segments D"B, CD and proceeding

from the pair D"B, CD        to the pair D"B, B'D = CD − D"B

as in Figure 10. Once again,

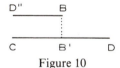

Figure 10

the common measures of D"B and CD

are precisely the same as

the common measures of D"B and B'D

so we can repeat the process once more with the pair of segments D"B, B'D (Figure 11). At this stage in our particular example we discover that

B'D *measures* D"B *exactly*

(in fact just once), thus

B'D *is the greatest common measure of the segments* D"B *and* B'D,

Figure 11

and, since the common measures remain the same at each stage of the above process, we may therefore conclude that

B′D *is the greatest common measure of the original pair of segments* AB, CD.

In fact this method will always produce the greatest common measure of two segments AB, CD, provided only that they have a common measure. We should not, of course, assume this on the basis of one particular example—even though you may have observed that the *method* is not dependent on the particular example we considered (the segments get shorter at each stage, and since they cannot ever become shorter than the common measure we are told exists, the process must stop—see Exercise 7). The formal proof that this method always works is a classic example of mathematical induction.

**Theorem 2.3.** *If two given segments* AB, CD *have an* (unknown) *common measure* MN, *then the above procedure will always produce the greatest common measure* XY *of* AB, CD *in finitely many steps.*
*If* $AB : MN = m$ *and* $CD : MN = n$, *and if* $h = hcf(m, n)$, *then* $XY : MN = h$.

PROOF. We use induction on the (unknown!) number $m + n$.
*Induction basis*: Suppose $m + n = 2$ (that is, $m = n = 1$). Then $AB = MN = CD$ and the above procedure produces the greatest common measure straight away: thus the first part of the assertion is true in this case. Moreover $XY = MN$ and $hcf(m, n) = 1$, so the second assertion is also true.
*Induction step*: Suppose that whenever $m + n < k$, both parts of the assertion are correct. Now consider the case where $m + n = k$. If $AB = CD$, then the above procedure produces the greatest common divisor straight away and both parts of the assertion are true as before.

If $AB \neq CD$, then one of the segments must be longer than the other, so we may assume that $AB > CD$. After one stage of our procedure (Figure 12), we obtain segments D′B, CD where $D′B : MN = (m - n) : 1$ and

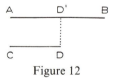

Figure 12

$CD : MN = n : 1$ *as before. Since* $(m - n) + n = m < m + n = k$, *both parts of the assertion are correct for the pair of segments* D′B, CD (*by induction*). Thus the above procedure will produce the greatest common measure of D′B and CD in finitely many steps. If the number of steps required by this procedure is $N$, then counting the previous step (from the pair AB, CD to the pair D'B, CD) we see that the same procedure produces the greatest common measure of AB and CD in $1 + N$ steps: thus the first part of the theorem is true for the pair AB, CD. Moreover, the second part of the assertion is true for the pair D'B, CD by induction, and the greatest common measure XY of D'B and CD is precisely the same as the greatest common measure of AB and CD; since $hcf(m - n, n) = hcf(m, n) = h$, the second part of the theorem is also true for the pair AB, CD.

This completes the induction step, so the theorem is true for all values of $m + n$.                                                              □

In the next section we shall see that if AB and CD correspond to the side and diagonal of a square, then the procedure described above *goes on for ever*: that is, it does *not* terminate after a finite number of steps as it would have to do if AB and CD had a common measure.

EXERCISES

1. Consider the game of Euclid with the initial pair (13, 18) and player A to start. If we list all possible positions in the following way, then it is easier to see which moves are most significant:

(13, 18)
(13, 5) → (13, 5)
    (8, 5)
    (3, 5) → (3, 5)
        (3, 2) → (3, 2)
           (1, 2) → (1, 2)
           (1, 1)
           (1, 0)

*"Player A has no choice, so must move to* (13, 5).
*Player B now has a critical choice : if B moves to* (3, 5) *then A wins; but if B moves to* (8, 5) *then B wins. ..."*
Analyse the games with initial pairs

(41, 23),      (43, 34),      (55, 34)

in the same way to see whether A (or B) can definitely win, and what moves she/he should make to do so.

2. Given a list of all possible positions as in Exercise 1, a move to the bottom pair in any column is called an *ultimate* move, and a move to the next bottom pair in any column is called a *penultimate* move. In the (13, 18) game B makes both ultimate and penultimate moves, and wins; in the (55, 34) game B makes only ultimate moves (she/he has no choice!) and still wins. What is the *simplest* initial pair (A to start), for which B makes only penultimate moves and still wins?

3. In the game with initial pair (13, 18) and A to start, B has a choice between the penultimate move (13, 5) → (8, 5) and the ultimate move (13, 5) → (3, 5): the penultimate move is in this case victorious, while the ultimate move is fatal.
   Show that whenever a player has a choice of two or more moves, precisely one

of the possible moves is (potentially) victorious and all others are (potentially) fatal.

Conclude that a move which is neither ultimate nor penultimate (e.g. $(18, 5) \rightarrow (13, 5)$) is *always* fatal.

Thus the key to the game of Euclid lies in knowing, when confronted with a choice,

when to choose the ultimate move $\rightarrow (a, b)$, where $1 < a/b < 2$, and

when to choose the penultimate move $\rightarrow (a + b, b)$.

4. Use the Euclidean method to work out the following:

| | |
|---|---|
| (i) $hcf(196, 91)$ | (vi) $hcf(5661, 5291)$ |
| (ii) $hcf(297, 140)$ | (vii) $hcf(5291, 4292)$ |
| (iii) $hcf(1632, 833)$ | (viii) $hcf(7576, 6591)$ |
| (iv) $hcf(2744, 675)$ | (ix) $hcf(325104, 221946)$ |
| (v) $hcf(41275, 4572)$ | (x) $hcf(3452, 1837)$ |

5. You may have already come across the sequence $u_1, u_2, u_3, \ldots$ of Fibonacci numbers:

$$1, 1, 2, 3, 5, 8, 13, 21, 34, 55, 89, 144, \ldots$$

whose first two terms are $u_1 = 1$ and $u_2 = 1$, with every other term $u_n$ $(n \geq 3)$ equal to the sum of the previous two terms: thus

$$u_1 = 1, \quad u_2 = 1, \quad u_n = u_{n-1} + u_{n-2} (n \geq 3).$$

(i) It is easy to check for the first few terms that
$hcf(1, 1) = hcf(1, 2) = hcf(2, 3) = hcf(3, 5) = hcf(5, 8) = 1.$

Prove (by induction on $n$) that $hcf(u_n, u_{n+1}) = 1$ for every $n \geq 1$.

(ii) Which terms of the Fibonacci sequence are divisible by 2? Make an intelligent guess, and try to prove it (by induction again). Which terms of the Fibonacci sequence are divisible by 3? Make an intelligent guess, and prove it. Which terms are divisible by 5? Generalise. [Hint: Show that

$$u_n = u_{n-1} + u_{n-2} = u_{n-2} + 2u_{n-3} = 2u_{n-3} + 3u_{n-4} = \text{etc.}]$$

(iii) Which terms of the Fibonacci sequence are divisible by 4? Which terms are divisible by 6? By 7? By 8? By 9? Try to find a general rule which will allow you to predict which is the first term divisible by $n$. Which terms thereafter are also divisible by $n$?

6. In the text when proving that $hcf(c, 0) = c$; we stated that "clearly $c$ is the largest factor of $c$." This is true for a winning pair $(c, 0)$ in the game of Euclid, but only because in this case $c$ *cannot be equal to* 0. Suppose $c = 0$, does there exist a whole number which we could call $hcf(c, 0)$?

7. (i) Consider two line segments AB, CD, where AB is 3 *kilometres* long and CD is 49.5 *centimetres* long. Which of the following lengths correspond to common measures of AB and CD?

| | |
|---|---|
| (a) 1 metre | (b) .5 metres |
| (c) 1 centimetre | (d) .5 cms |
| (e) .1 cms | (f) .05 cms |
| (g) .01 cms | (h) .005 cms |
| (i) .001 cms | (j) .0005 cms |

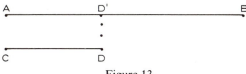

Figure 13

Does there exist a *smallest* common measure of the segments AB and CD?

(ii) On page 44 we observed that if $D'B = AB - CD$ (Figure 13), then the common measures of AB and CD are precisely the same as the common measures of D′B and CD. We then remarked: "in particular, the *greatest* common measure of AB and CD is precisely the same as the *greatest* common measure of D′B and CD." Since you now realise that there is no such thing as a *smallest* common measure, you should explain why we can be so sure about the existence of a *greatest* common measure. [HINT: On page 44 we had no right to be so sure: the existence of a *greatest* common measure is strictly speaking a consequence of Theorem 2.3.]

8. In the text we have stuck to line segments the whole time, but there is a much more graphic way of presenting the method of comparing line segments AB and CD which starts out from *the rectangle spanned by* AB and CD.
(i) (a) Take a sheet of graph paper. Choose two segments AB, CD with $B = C$, and such that the two segments AB, CD span a rectangle ABDZ (Figure 14).

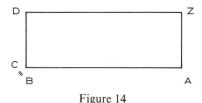

Figure 14

(b) Instead of simply subtracting CD from AB as many times as possible, cut off CD by CD squares from the rectangle ABDZ as many times as possible (Figure 15), leaving the rectangle $A_1BDZ_1$.

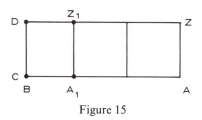

Figure 15

(c) Next, instead of simply subtracting $A_1B$ from CD as many times as possible, cut off $A_1B$ by $A_1B$ squares from the rectangle $A_1BDZ_1$ as many times as possible (Figure 16), leaving the rectangle $A_1BD_1Z_2$.
(d) Next, cut off $CD_1$ by $CD_1$ squares as many times as possible (Figure 17), leaving the rectangle $A_2BD_1Z_3$.
(e) Continue until one arrives either at a rectangle $A_iBD_iZ_{2i}$ which can be dissected into a whole number of $CD_i$ by $CD_i$ squares (in which case $CD_i$ is the

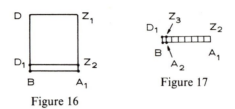

Figure 16                    Figure 17

required greatest common measure of the segments AB and CD), or at a rectangle $A_{i+1}BD_iZ_{2i+1}$ which can be dissected into a whole number of $A_{i+1}B$ by
$A_{i+1}B$ squares (in which case $A_{i+1}B$ is the required greatest common measure).

(ii) Apply the method of part (i)

(a) when AB = 6.8, CD = 4.2:
(b) when AB = 9.1, CD = 14.3.

CHAPTER II.4

# Sides and Diagonals of
# Regular Polygons

Let us now apply the procedure developed in the previous chapter to try to
find a common measure for the diagonal AC and the side BC of a square
ABCD.

First choose the point B′ on AC such that

$$AB' = AB \ (= BC)$$

Then

the common measures of AC and BC

are precisely the same as

the common measures of B′C ($=AC - BC$) and BC (Figure 18).

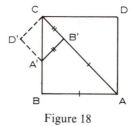

Figure 18

Let the perpendicular to AC at B′ meet BC in A′. Then

B′C = B′A′ (since the triangle A′B′C is isosceles),

and

B′A′ = BA′ (by Exercise 1(i)).

Thus B′C = BA′, so

the common measures of B′C and BC

are precisely the same as

the common measures of B′C and A′C ($=BC - B'C$).

But now B′C and A′C are just the side and diagonal of the square A′B′CD′.

Thus we have proved that

> the common measures of the *diagonal* AC and the *side* BC of the
> large square ABCD

are precisely the same as

> the common measures of the *side* B'C and the *diagonal* A'C of the
> smaller square A'B'CD'.

We can therefore repeat the above process—starting this time with the
side B'C and the diagonal A'C of the smaller square A'B'CD'.
Choose the point B'' on A'C such that

$$A'B' = A'B''\ (=B'C).$$

Then

> the common measures of B'C and A'C

are precisely the same as

> the common measures of B'C and B''C ($=A'C - B'C$) (Figure 19).

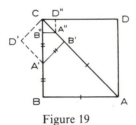

Figure 19

Let the perpendicular to A'C at B'' meet B'C in A''. Then

$$B''C = B''A''\ \text{(since the triangle } A''B''C \text{ is isosceles)},$$

and

$$B''A'' = B'A''\ \text{(by Exercise 1(i))}.$$

Thus $B''C = B'A''$, so

> the common measures of B'C and B''C

are precisely the same as

> the common measures of A''C ($=B'C - B''C$) and B''C.

But now A''C and B''C are just the *diagonal* and *side* of the square
A''B''CD''.
And so we could continue indefinitely—passing from the *diagonal* and
*side* of one square, to the *side* and *diagonal* of a smaller square, to the
*diagonal* and *side* of a still smaller square.

But what, you may ask, has all this to do with the assertion that the diagonal AC and the side BC of the original square ABCD have no common measure? In answering we may argue in either of the following two ways—the first is *indirect*, the second is (in some sense) *direct* and *constructive*.

First the indirect proof: we have shown that, when we apply our general procedure for finding common measures to the diagonal AC and side BC of the square ABCD, the process *goes on for ever*. However, Theorem 2.3 states that

*if* AC and BC have a common measure
*then* the process must terminate *after finitely many steps.*

This contradiction can only arise because AC and BC do not in fact have a common measure.

Second, the direct proof: we interpret our general procedure as a *machine for constructing* the (greatest) common measure XY of any two segments, in the sense that

*if* the two segments have a common measure
*then* our procedure will construct the greatest common measure.

The application of this foolproof machine to the diagonal AC and side BC of the square ABCD in Figure 20 can then be interpreted as showing that a

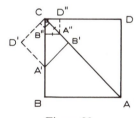

Figure 20

greatest common measure XY of AC and BC would have to satisfy

$XY \leq BC$ (since XY is a common measure of AC and BC);

$XY \leq B'C < (1/2) \cdot BC$ (since XY is a common measure of B'C and A'C, and $B'C < (1/2)BC$ by Exercise 1(ii));

$XY \leq B''C < (1/2)^2 \cdot BC$ (since XY is a common measure of B''C and A''C, and $B''C < (1/2)B'C < (1/2)^2 \cdot BC$ by Exercise 1(ii)).

Continuing in this way we see that the common measure XY we are trying to construct must satisfy

$$XY < (1/2)^n \cdot BC \qquad \textit{for every whole number } n.$$

Since this cannot possibly be true, our direct, constructive procedure produces a negative conclusion in this case: AC and BC have no common measure.

This phenomenon (of *diagonal* and *side* in some larger figure being related in this way to *side* and *diagonal* in some smaller figure) is not confined to the square: the regular pentagon exemplifies this in a particularly pleasing way.

Consider the common measures of the diagonal AD and side BC of the regular pentagon ABCDE (Figure 21). BC is parallel to AD, and CD is parallel to BE, so BCDB′ must be a parallelogram; in particular BC = DB′.

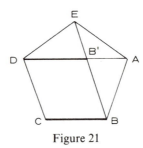

Figure 21

Thus

the common measures of BC and AD

are precisely the same as

the common measures of BC and AB′ ( = AD − BC) (Figure 22).

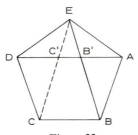

Figure 22

Since AB is parallel to CE, ABCC′ must also be a parallelogram; in particular BC = AC′, so B′C′ = BC − AB′. Thus

the common measures of BC and AB′

are precisely the same as

the common measures of B′C′ and AB′.

If we now draw in the remaining two diagonals BD and AC, we see that the five diagonals enclose a regular pentagon A'B'C'D'E' (Figure 23). AD is

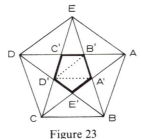

Figure 23

parallel to A'D', and AC is parallel to B'D', so AB'D'A' must be a parallelogram; in particular AB' = A'D'. Thus our previous two steps,

    from BC, AD        to BC, AB'

and   from BC, AB'      to B'C', AB' = A'D'

can be interpreted as showing that

the common measures of the side BC and diagonal AD of the large pentagon ABCDE

are precisely the same as

the common measures of the side B'C' and diagonal A'D' of the smaller pentagon A'B'C'D'E'.

We can obviously continue in this way, passing first from the large pentagon ABCDE to the smaller pentagon A'B'C'D'E',[1] then from the smaller pentagon A'B'C'D'E' to the still smaller pentagon A"B"C"D"E" and so on indefinitely (Figure 24).

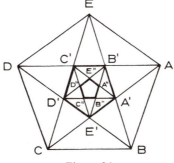

Figure 24

---

[1] Observe that we are *not* simply proving that the *ratio* of the side to the diagonal of the larger pentagon ABCDE is equal to the *ratio* of the side to the diagonal in the smaller pentagon A'B'C'D'E'—this is obvious, since the figures ABCDE and A'B'C'D'E' are similar. What we have proved is that *any unit which measures both* AD *and* BC *will also measure* A'D' *and* B'C'.

As for the square we may conclude that AD and BC have no common measure in either of two ways. Firstly the process continues indefinitely, whereas Theorem 2.3 states that

*if* AD and BC have a common measure
*then* the process must terminate *after finitely many steps*,

and so we have a contradiction unless AD, BC have no common measure. Alternatively, interpreting our general procedure as a machine for constructing a (greatest) common measure XY of AD and BC, we conclude (Exercise 2) that

$$XY \le BC, \quad XY \le B'C' < \frac{1}{2} \cdot BC, \quad XY \le B''C'' < \frac{1}{2^2} \cdot BC, \ldots$$

and so

$$XY < \frac{1}{2^n} \cdot BC \quad \textit{for every whole number n;}$$

since this cannot possibly be true, our constructive procedure registers a negative result: AD and BC have no common measure.

EXERCISES

1. Given a square ABCD and a point B' on AC such that AB = AB'. Draw the perpendicular to AC at B', meeting BC in A' (Figure 25).

Figure 25

   (i) Prove that BA' = B'A'.

   (ii) Prove that B'C < (1/2) · BC.

2. Given a regular pentagon ABCDE, with diagonals enclosing the smaller pentagon A'B'C'D'E', prove that B'C' < (1/2) · BC (Figure 26).

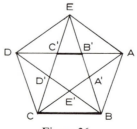

Figure 26

3. Given a regular hexagon ABCDEF, the six diagonals AC, BD, CE, DF, EA, FB enclose a regular hexagon A′B′C′D′E′F′ (Figure 27).

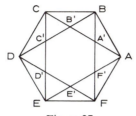

Figure 27

(i) Show that the diagonal A′C′ of the smaller hexagon is equal to the side BC of the larger hexagon.

(ii) Show that the side A′B′ of the smaller hexagon is precisely one third the diagonal AC of the larger hexagon.

(iii) Conclude that A′B′ ≠ AC − BC (so the argument which worked for squares and regular pentagons does not seem to carry over to regular hexagons).

4. Given a regular octagon ABCDEFGH, the diagonals AD, BE, ..., HC enclose a regular octagon A′B′C′D′E′F′G′H′ (Figure 28).

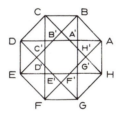

Figure 28

(i) Show that the diagonal H′C′ of the smaller octagon is equal to the side BC of the larger octagon.

(ii) Show that AA′ = BC and B′D = BC.

(iii) Deduce that the common measures of the side BC and diagonal AD of the larger octagon are precisely the same as the common measures of the diagonal H′C′ and the side A′B′ of the smaller octagon.

(iv) Hence conclude that BC and AD have no common measure.

5. Given the regular 7-gon ABCDEFG, the diagonals AD, BE, ..., GC enclose a smaller 7-gon A′B′C′D′E′F′G′.

(i) Show that BC = B′A and BC = DC′. Conclude that the common measures of the side BC and diagonal AD of the larger 7-gon are precisely the same as the common measures of BC and the side B′C′ of the smaller 7-gon.

(ii) If CF meets AD in X and BG meets AD in Y show that BC = YX, so we may work with XY, B′C′ in place of BC, B′C′.

(iii) Prove that A′D′ = YB′.

(iv) Hence show that our general process does *not* yield the conclusion that the common measures of the side BC and diagonal of the larger 7-gon are precisely the same as the common measures of the diagonal A′D′ and side B′C′ of the smaller 7-gon (Figure 29).

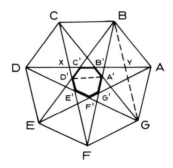

Figure 29

# Numbers and Arithmetic—
# A Quick Review

"Can you do addition?" the White Queen asked.
"What's one and one and one and one and one
and one and one and one and one and one?"
"I don't know," said Alice. "I lost count."

Lewis Carroll, *Through the Looking Glass*

Ein Paar Zeitungen
zwei Parteizungen,
Zwei Paar Zeitungen
vier Parteizungen, usw.

C. Morgenstern, *Arithmetische Progression*

Some mathematical ideas seem as subtle and difficult today as when they were first introduced; others, which could once upon a time be handled only by those with a very extensive training, now seem entirely elementary to us. For example, addition and multiplication of whole numbers was once a skilled occupation, but is today taught to all children.

Such changes in *apparent* difficulty are very often due to the invention of an *improved notation* which is better suited to the task at hand, and there are two examples which are of particular significance for this book. The most spectacular *elementary* example of improved notation having such a dramatic effect is *the representation of numbers in terms of a fixed number base*: we usually use base ten and the digits 0, 1, 2, 3, 4, 5, 6, 7, 8, 9, but the principle is exactly the same for any base. The most spectacular *advanced* example is the notation introduced by Leibniz in his differential calculus. It is one of the characteristics of a really effective notation that we use it without worrying too much about what exactly it represents. We perform multiplications or long divisions like a machine; if we try simultaneously to calculate *and* to think about what the symbols really represent, or why the method works, we can get into a terrible muddle. In the next few pages we shall begin to reflect on *numbers and their representation* in the decimal system. What we have to say at this stage is very elementary, and much of it will be more or less familiar; but you are nevertheless advised to read it *carefully*—though you may find that you can read it fairly quickly. In the next few chapters we shall move on to study infinite decimals; and in

Chapter II.13 we shall introduce a less familiar but very important way of representing numbers, which is essentially a numerical version of Euclid's method for constructing common measures, and which gives rise to infinite processes which are both intriguing and highly instructive; (those who remained sceptical about the assertion that the method of the previous chapter was *direct* and constructive will hopefully be convinced by the numerical version).

That the whole numbers *exist* in some sense, is a fact which is completely independent of the way we choose to write them down. The crudest kind of representation might be to introduce a separate stroke for each extra unit—thus

| || ||| |||| ||||| |||||| ||||||| |||||||| ||||||||| |||||||||| ||||||||||| ...

But beyond a certain point (around |||| or |||||) this representation ceases to convey any real information *at a glance*. This changes dramatically if the strokes are *grouped* in some agreed way, as in the familiar method of keeping a tally:

| || ||| |||| ⊥⊥⊥ ⊥⊥⊥| ⊥⊥⊥|| ⊥⊥⊥||| ⊥⊥⊥|||| ⊥⊥⊥⊥⊥ ⊥⊥⊥⊥⊥| ...

Another approach to the problem of representing numbers makes use of particular pictures or signs—restricting attention to those numbers which actually arise in everyday life.[1] Cultures whose written language is based on an alphabet may naturally use the commonly accepted sequence of the letters of the alphabet to correspond to the natural sequence of numbers, as we find in the ancient Greek *alphabetic* numerals:

$\alpha$   $\beta$   $\gamma$   $\delta$   $\varepsilon$   $\varsigma$   $\zeta$   $\eta$   $\theta$   $\iota$ ...

The familiar Roman numerals

I   II   III   IV   V   VI   VII   VIII   IX   X   XI ...

seem much closer to tallying, though they make use of two clever abbrevi-

---

[1] Our own *number names*

    one   two   three   four   five   six   seven   eight   nine   ten   eleven ...

are restricted in precisely this way. Unlike the decimal representation of numbers, which can in principle be extended indefinitely on the basis of the ten digits 0, 1, 2, 3, 4, 5, 6, 7, 8, 9, alone, number names are not generated automatically, but have to be *invented* as and when usage requires it: witness the physicists' invention of terms like terawatt (tera $= 10^{12}$), gigaelectronvolt (giga $= 10^9$), megaton (mega $= 10^6$), kilogram (kilo $= 10^3$), millimetre (milli $= 10^{-3}$), microhm (micro $= 10^{-6}$), picofarad (pico $= 10^{-12}$), .... The prefixes

    ..., tera-, giga-, mega-, kilo-, milli-, micro-, pico-, ...

form an internationally recognised verbal 'number system' (mega · mega = tera, etc.), which differs from ordinary powers of ten in that new names have to be invented from time to time whenever larger or smaller number names are unexpectedly needed.

ations: firstly, bundles of strokes are replaced by a single symbol (thus ⊥⊦⊦
becomes V); and secondly, a convention is introduced whereby IV denotes
"I before V" (that is, *four*).

For *counting small numbers* several of the above mentioned systems are
at least as good as our own

$$1 \quad 2 \quad 3 \quad 4 \quad 5 \quad 6 \quad 7 \quad 8 \quad 9 \quad 10 \quad 11 \ldots.$$

It is even arguable that they are *all more reasonable* than our own decimal
system, in so far as one can at least see some *logic* behind their sequence of
symbols. In the first two examples the logic is clear. In the third example
the logic consists of a familiar convention borrowed from the alphabet. The
Roman numerals build in a very natural way by making repeated use of the
two clever abbreviations mentioned, with new symbols like V, X, L, C, D,
M, etc. being introduced only when they are clearly needed. In contrast our
own system begins with *ten distinct abstract symbols*! And, with the exception of 1 whose *one-ness* is apparent, each symbol seems totally unrelated to
the *number* it is supposed to represent: there is, for example, a strong whiff
of *two-ness*, or *second-ness*, about

$$||, \quad \text{or} \quad \text{II}, \quad \text{or (for a Greek) } \beta,$$

which is totally missing from the abstract squiggle 2.

But the first striking advantage of our own system shows up as soon as
one tries to count larger numbers. This advantage stems from the simple,
repetitive way in which we manipulate the ten fundamental symbols, or
digits. Among other things, in our own system this simple, repetitive counting process suggests in a natural way the totally astonishing idea that *the
sequence of whole numbers is endless*, whereas in other systems such an idea
is not at all obvious.

The second really striking advantage of our own system emerges when
one comes to do elementary arithmetic. Though one can manage addition
in most systems—at least for small numbers—subtraction, multiplication
and division are usually gruesome in the extreme (Exercise 2). In contrast,
representing a number in terms of the powers of a fixed base—whether that
base be the familiar *base 10* or some other—reduces all these operations to
the implementation of the processes we know so well. Whatever base is
used, each number is expressed in a form analogous to a polynomial

$$5931 = 5 \cdot 10^3 + 9 \cdot 10^2 + 3 \cdot 10 + 1 \qquad 5x^3 + 9x^2 + 3x + 1$$

whence addition and subtraction can be carried out simply by *collecting
together like terms*:

$$
\begin{aligned}
5931 &= 5.10^3 + \phantom{0}9.10^2 + 3.10 + 1 \\
+\,1248 &= 1.10^3 + \phantom{0}2.10^2 + 4.10 + 8 \\
\hline
\text{****} \quad &\phantom{=}\ 6.10^3 + 11.10^2 + 7.10 + 9
\end{aligned}
$$

$$5931 = 5.10^3 + 9.10^2 + 3.10 + 1$$
$$-1248 = 1.10^3 + 2.10^2 + 4.10 + 8$$

$$****\quad 4.10^3 + 7.10^2 + (-1).10 + (-7)$$

though some adjustment is required if we are to stick to the convention that only the digits 0–9 should appear as coefficients. Thus

$$11.10^2 = (10 + 1).10^2 = 1.10^3 + 1.10^2$$

and

$$(-1).10 + (-7) = (-1).10^2 + 8.10 + 3$$

so

$$5931 + 1248 = 6.10^3 + 11.10^2 + 7.10 + 9$$
$$= 7.10^3 + \quad 1.10^2 + 7.10 + 9$$
$$= 7179$$

$$5931 - 1248 = 4.10^3 + 7.10^2 + (-1).10 + (-7)$$
$$= 4.10^3 + 6.10^2 + 8.10 + 3$$
$$= 4683.$$

Multiplication can also be carried out by treating the decimal representation as analogous to a polynomial, and *collecting together like terms*:

$(x + 1)^2 = x^2 + 2x + 1$
$11^2 = (1.10 + 1)^2 = 1.10^2 + 2.10 + 1$
$\quad\quad\quad\quad\quad\quad = 121$

$(x + 1)^3 = x^3 + 3x^2 + 3x + 1$
$11^3 = (1.10 + 1)^3 = 1.10^3 + 3.10^2 + 3.10 + 1$
$\quad\quad\quad\quad\quad\quad = 1331$

$(x + 1)^4 = x^4 + 4x^3 + 6x^2 + 4x + 1$
$11^4 = (1.10 + 1)^4 = 1.10^4 + 4.10^3 + 6.10^2 + 4.10 + 1$
$\quad\quad\quad\quad\quad\quad = 14641$

$(x + 1)^5 = x^5 + 5x^4 + 10x^3 + 10x^2 + 5x + 1$
$11^5 = (1.10 + 1)^5 = 1.10^5 + 5.10^4 + 10.10^3 + 10.10^2 + 5.10 + 1$
$\quad\quad\quad\quad\quad\quad = 1.10^5 + 6.10^4 + 1.10^3 + 0.10^2 + 5.10 + 1$
$\quad\quad\quad\quad\quad\quad = 161051$

(where the final example has been adjusted to conform to the convention that only the digits 0–9 appear as coefficients).

Division is carried out in essentially the same simple minded way, though the standard procedures do not assist us in thinking what exactly we are doing. For example, the sequence of steps we go through when we work out

$$17 \overline{)3\ 8\ 5\ 9}$$
$$\frac{\cdot\ \cdot}{\cdot\ 5}$$
$$\frac{\cdot\ \cdot}{\cdot\ \cdot\ 9}$$

can be written rather differently as

$$3859 = 34 \cdot 10^2 + 459$$

$$= 17 \cdot (2 \cdot 10^2) + 459$$

$$459 = 34 \cdot 10 + 119$$

$$= 17 \cdot (2 \cdot 10) + 119$$

$$119 = 17.7$$

whence

$$3859 = 17(2 \cdot 10^2 + 2 \cdot 10 + 7)$$

so

$$\frac{3859}{17} = 227$$

But our notation is not only helpful in simplifying arithmetic, and in suggesting so naturally the existence of an endless sequence of counting numbers: it has one further consequence which is more significant than either of these. *It can be extended beyond the decimal point in a completely natural way, to yield a representation of fractional parts of the unit; and this representation is in complete harmony with the decimal representation of whole numbers.*

As we have already observed, such convenient flexibility can be a mixed blessing. Too often, for example, a child may be told

"$50 + 20 = 70$   *because*   $5 + 2 = 7$."

Unfortunately the beginner is unlikely to see any logical connection whatsoever between the two relations. Of course, our *final* understanding of the first relation ($50 + 20 = 70$) would not be complete if we did not experience the symbols *50* and *20* as *5 lots of ten* and *2 lots of ten* respectively, and so interpret their sum, *50 + 20*, as *7 lots of ten;* but this all takes time.[2] For us

---

[2] In a similar way our own understanding of the relation $45 + 18 = 63$ includes our experiencing the symbols *45* and *18* as *5 lots of nine* and *2 lots of nine* respectively, though a beginner would see no logical connection between $45 + 18 = 63$ and $5 + 2 = 7$.

the experience of the standard addition procedure

$$\begin{array}{r} 50 \\ +\,20 \\ \hline 70 \end{array}$$

with its ritual incantation "$0 + 0 = 0, \; 5 + 2 = 7$", is so strong that we mistake it for an *explanation*.

When introducing decimals it is just as tempting to suggest that

$$\left. \begin{array}{l} .18 + \;\; .25 = \;\; .43 \\ \text{and} \;\; .018 + .025 = .043 \end{array} \right\} \quad because \quad 18 + 25 = 43;$$

but the possibility of confusion is now much greater. For though

$$.18 \times .025 = \underline{\hspace{1.5cm}} ? \quad \text{and} \quad \frac{.18}{.025} = \underline{\hspace{1.5cm}} ?$$

*can* be solved by first working out

$$18 \times 25 = 450 \quad \text{and} \quad \frac{18}{25} = .72,$$

and then applying some rule or other to position the decimal point "correctly," the corresponding addition and subtraction problems

$$.18 + .025 = \underline{\hspace{1.5cm}} ? \quad \text{and} \quad .18 - .025 = \underline{\hspace{1.5cm}} ?$$

have nothing whatever to do with $18 \pm 25$.

Our understanding of decimal arithmetic is only hampered by such rules. To appreciate decimal arithmetic we must come to grips with the meaning of the decimal representation of fractions, which is an extension of the "polynomial" form for whole numbers. But whereas

*whole numbers are expressed in terms of the powers of 10*

in a manner analogous to polynomials in $x$,

*decimal fractions are expressed in terms of the powers of 1/10*

in a manner analogous to polynomials in $1/x$. Thus

$$.18 = 1 \cdot \frac{1}{10} + 8 \cdot \frac{1}{10^2} \qquad\qquad 1 \cdot \frac{1}{x} + 8 \cdot \frac{1}{x^2}$$

$$.025 = 0 \cdot \frac{1}{10} + 2 \cdot \frac{1}{10^2} + 5 \cdot \frac{1}{10^3} \qquad 0 \cdot \frac{1}{x} + 2 \cdot \frac{1}{x^2} + 5 \cdot \frac{1}{x^3}$$

so addition and subtraction can be carried out, as for polynomials, by *collecting together like terms :*

$$.18 = 1 \cdot \frac{1}{10} + 8 \cdot \frac{1}{10^2}$$

$+$

$$.025 = 0 \cdot \frac{1}{10} + 2 \cdot \frac{1}{10^2} + 5 \cdot \frac{1}{10^3}$$

$$1 \cdot \frac{1}{10} + 10 \cdot \frac{1}{10^2} + 5 \cdot \frac{1}{10^3}$$

$$.18 = 1 \cdot \frac{1}{10} + 8 \cdot \frac{1}{10^2}$$

$-$

$$.025 = 0 \cdot \frac{1}{10} + 2 \cdot \frac{1}{10^2} + 5 \cdot \frac{1}{10^3}$$

$$1 \cdot \frac{1}{10} + 6 \cdot \frac{1}{10^2} + (-5) \cdot \frac{1}{10^3}$$

The results must then be adjusted to conform to the convention that only the digits 0–9 should appear as coefficients: thus

$$10 \cdot \frac{1}{10^2} = 1 \cdot \frac{1}{10} \quad \text{and} \quad (-5) \cdot \frac{1}{10^3} = (5 - 10) \cdot \frac{1}{10^3} = 5 \cdot \frac{1}{10^3} + (-1) \cdot \frac{1}{10^2}$$

so we obtain

$$.18 + .025 = 2 \cdot \frac{1}{10} + 0 \cdot \frac{1}{10^2} + 5 \cdot \frac{1}{10^3}$$

$$= .205$$

and

$$.18 - .025 = 1 \cdot \frac{1}{10} + 5 \cdot \frac{1}{10^2} + 5 \cdot \frac{1}{10^3}$$

$$= .155$$

Multiplication is carried out in exactly the same spirit: if presented with

$$.18 \times .025 = \left(1 \cdot \frac{1}{10} + 8 \cdot \frac{1}{10^2}\right) \times \left(0 \cdot \frac{1}{10} + 2 \cdot \frac{1}{10^2} + 5 \cdot \frac{1}{10^3}\right)$$

first multiply out the brackets as for polynomials, and then *collect together like terms*:

$$= 2 \cdot \frac{1}{10^3} + (8 \times 2 + 1 \times 5) \cdot \frac{1}{10^4} + (8 \times 5) \cdot \frac{1}{10^5} \cdot$$

This result must then be suitably adjusted, using

$$(8 \times 5) \cdot \frac{1}{10^5} = 4 \cdot \frac{1}{10^4}$$

and $\qquad (8 \times 2 + 1 \times 5) \cdot \frac{1}{10^4} + 4 \cdot \frac{1}{10^4} = 2 \cdot \frac{1}{10^3} + 5 \cdot \frac{1}{10^4},$

to obtain

$$.18 \times .025 = 2 \cdot \frac{1}{10^3} + 2 \cdot \frac{1}{10^3} + 5 \cdot \frac{1}{10^4}$$

$$= .0045.$$

Division is admittedly most easily understood if we openly remove the difficulties created by the decimal points, and rewrite the problem as follows:

$$\frac{.18}{.025} = \frac{.18}{.025} \times 1 = \frac{.18}{.025} \times \frac{10^3}{10^3}$$

$$= \frac{\left(1 \cdot \frac{1}{10} + 8 \cdot \frac{1}{10^2}\right) \cdot 10^3}{\left(2 \cdot \frac{1}{10^2} + 5 \cdot \frac{1}{10^3}\right) \cdot 10^3} = \frac{1 \cdot 10^2 + 8 \cdot 10}{2 \cdot 10 + 5}$$

$$= \frac{180}{25}.$$

The sequence of steps we then go through in the usual division procedure

$$25 \overline{\smash{\big)}\,1\ 8\ 0}$$

can be expressed in the form

$$180 = 25 \times 7 + 5$$

$$5 = 50 \times \frac{1}{10}$$

$$= 25 \times \left(2 \cdot \frac{1}{10}\right)$$

so

$$180 = 25 \times \left(7 + 2 \cdot \frac{1}{10}\right)$$

whence

$$\frac{.18}{.025} = \frac{180}{25} = 7 + 2 \cdot \frac{1}{10} = 7.2.$$

So far then, so good. But what about all those numbers like 1/3 and 8/70 whose decimal representation goes on forever[3]? Can we extend the usual addition procedure to cope with expressions like the following?

$$+ \begin{array}{l} .33333 \dots \\ .11428571428571 \dots \end{array}$$

Remember that, in order to keep control of *carrying*, we usually start with the *right hand column!* And even if we discover some way of adding and subtracting these expressions, how can we possibly carry out multiplication and division? If you have never thought of decimal arithmetic like this before, these questions may come as something of a shock. And though they *can* be answered, they suggest even more serious questions. For example, once we decide to accept tailless monsters like

$$.\dot{3} = .3333 \dots \quad \text{and} \quad .1\dot{1}4285\dot{7} = .11428571428571428 \dots$$

simply because they arise from fractions, should we therefore accept *all* tailless monsters? In what sense, if any, do

$$.989989998999989999989 \dots$$

and

$$.123456789101112131415 \dots$$

represent *numbers?* If they do, can they be added, subtracted, multiplied, and divided like ordinary numbers? How should I begin to work out the sum, or product, of the two expressions above? We shall begin to answer these questions in the next few chapters, *but in order to appreciate the discussion in those chapters you should make a serious attempt at Exercise 4.*

### EXERCISES

1. Suppose you are devising your own notation to represent numbers, and you decide to use the letters of the English alphabet to represent the first twenty-six numbers ($a = 1$, $b = 2$, ..., $z = 26$). Invent two distinct systems for extending this notation so that it can be used to represent larger numbers: you are only allowed to use the symbols $a$ to $z$, but strings of symbols and repetitions are allowed.

---

[3] $1/3 = .\dot{3} = .3333 \dots$ and $8/70 = .1\dot{1}4285\dot{7} = .11428571428571 \dots$, where we use a standard notation—placing dots over the first and last digits of a repeating block.

Write the numbers

twenty-seven      fifty-three      seven hundred,

in each of your two systems.

2. Try to find an effective procedure for solving *Roman numeral arithmetic* problems: you must work wholly within the Roman numeral system, and are not allowed to translate the numbers into any other system.

(i) XXI + IX = _____      (iii) XIV × XLV = _____

(ii) XIII − IV = _____      (iv) MCC ÷ XL = _____

3. (i) Show that $x^i - 1$ always has $x - 1$ as a factor.

(ii) Let the number $N$ be represented as

$$a_n a_{n-1} \cdots a_0$$

in base $k$. Show that

if $a_n + a_{n-1} + \cdots + a_0$ is a multiple of $k - 1$,
then $N$ is also a multiple of $k - 1$.

[Hint: $N = a_n k^n + a_{n-1} k^{n-1} + \cdots + a_1 k + a_0$, and $a_i k^i = a_i(k^i - 1) + a_i$.]

(iii) Which of the following numbers (base 10) are divisible by 9?

72;   108;   162;   423;   441;   506;   979;   503621;   503622;   123456789.

(iv) Look again at your solution to (ii). Let $d$ be any number dividing $k - 1$. Show that

if $a_n + a_{n-1} + \cdots + a_0$ is a multiple of $d$,
then $N$ is also a multiple of $d$.

(v) Which of the following numbers (base 10) are multiples of 3?

1;   12;   123;   1234;   12345;   123456;   1234567;   12345678;   123456789.

(vi) Show that whenever $i$ is even, $x^i - 1$ has $x + 1$ as a factor, and that whenever $i$ is odd, $x^i + 1$ has $x + 1$ as a factor.

(vii) Let the number $N$ be represented as

$$a_n a_{n-1} \cdots a_0$$

in base $k$. Show that

if $a_n - a_{n-1} + \cdots + (-1)^n a_0$ is a multiple of $k + 1$,
then $N$ is also a multiple of $k + 1$.

Extend this result as in (iv) to the case where $d$ is any number dividing $N$, and $a_n - a_{n-1} + \cdots + (-1)^n a_0$ is a multiple of $d$.

(viii) Which of the following numbers (base 10) are divisible by 11?

154;   165;   176;   264;   275;   616;   737;   858;
1309;   11011;   910111213141516.

4. (i) Work out each of the following; in each case describe the method you have used.

$$\text{(a) } 173 - 162 = \underline{\hspace{1cm}} ; \qquad \text{(b) } 162 - 173 = \underline{\hspace{1cm}} .$$

(ii) Work out each of the following; in each case describe the method you have used. (All the decimals involved are endless decimals.)

(a)  $.111\ldots + .222\ldots = \underline{\hspace{1cm}} ;$

(b)  $.888\ldots + .333\ldots = \underline{\hspace{1cm}} ;$

(c)  $.\dot1\dot8 + .\dot2 = \underline{\hspace{1cm}} ;$

(d)  $.\dot3 + .\dot114285\dot7 = \underline{\hspace{1cm}} ;$

(e)  $.\dot1 + .989989998999989\ldots = \underline{\hspace{1cm}} ;$

(f)  $.989989998999989\ldots + .1234567891011121\ldots = \underline{\hspace{1cm}} ;$

(g)  $.\dot1 \times .\dot2 = \underline{\hspace{1cm}} ;$

(h)  $.\dot8 \times .\dot3 = \underline{\hspace{1cm}} ;$

(i)  $.\dot1\dot8 \times .\dot2 = \underline{\hspace{1cm}} ;$

(j)  $.\dot3 \times .\dot114285\dot7 = \underline{\hspace{1cm}} ;$

(k)  $.\dot1 \times .989989998999989\ldots = \underline{\hspace{1cm}} ;$

(l)  $.\dot1 \times .123456789101112\ldots = \underline{\hspace{1cm}} ;$

(m) $.989989998999989\ldots \times .123456789101112\ldots = \underline{\hspace{1cm}} .$

(iii) Work out each of the divisions

$$\frac{.\dot1}{.\dot2}, \quad \frac{.\dot8}{.\dot3}, \quad \frac{.\dot1\dot8}{.\dot2}, \quad \text{etc.}$$

corresponding to the parts (a)–(m) in (ii).

# Infinite Decimals (Part 1)

Those who introduce us to counting numbers usually go to great lengths both to invest these mathematical entities with *ordinary meaning* and to clarify the *mathematical idea* behind the usual base 10 representation of numbers (that is, the idea of *place value* corresponding to increasing powers of 10). Similar efforts are made when the time comes to introduce negative whole numbers, fractions, and the decimal representation of decimal fractions. But though the long division process, which we use to transform an ordinary fraction into a decimal fraction, frequently gives rise to *infinite decimals*, little if any time or effort is devoted either to investing these curious entities with ordinary meaning, or to clarifying the mathematical idea which justifies their representation as *never ending decimals*. Experience suggests that many undergraduates complete their studies of sequences, series, and limits in the calculus without ever realising the light they shed on infinite decimals. But since the very essence of the calculus lies in the careful use it makes of *infinite processes* to supplement the familiar processes of ordinary arithmetic, and since infinite decimals constitute the most familiar example of such infinite processes, it seems rather obvious that these should be the *very first* candidates for analysis: this is precisely the aim of the next few chapters.

Many fractions can be safely transformed into decimal form: for example

$$\frac{1}{2} = \frac{1}{2} \times \frac{5}{5} = \frac{5}{10} = .5; \quad \frac{1}{5} = \frac{1}{5} \times \frac{2}{2} = \frac{2}{10} = .2;$$

$$\frac{1}{8} = \frac{1}{8} \times \frac{125}{125} = \frac{125}{1000} = .125; \quad \frac{3}{40} = \frac{3}{40} \times \frac{25}{25} = \frac{75}{1000} = .075.$$

In general, given an ordinary fraction $a/b$, where $a$, $b$ are whole numbers, we simply look for a whole number $c$ such that

$$b \times c = 10^n \quad \text{for some } n;$$

in this way we obtain an expression for $a/b$ as a decimal fraction:

$$\frac{a}{b} = \frac{a}{b} \times \frac{c}{c} = \frac{a \times c}{10^n}$$

In practice we usually obtain the decimal representation of $a/b$ by *dividing b into a*: thus, instead of calculating the decimal representation of 3/40 strictly as we did above, we plough ahead with the long division

$$
\begin{array}{r}
.0\ .\ . \\
40\,\overline{\smash{\big)}\,3.0\,0\,0\,0\,0\,0\,0\ldots} \\
\underline{\cdot\ \cdot\ \cdot} \\
\cdot\ \cdot\ 0 \\
\underline{\cdot\ \cdot\ \cdot}
\end{array}
$$

using up as many 0's as we seem to need. Now this procedure is perfectly valid, and can be justified as follows:

$$
\frac{3}{40} = \frac{3}{40} \times \frac{10^3}{10^3} = \frac{3000}{40} \times \frac{1}{10^3}
$$

and

$$
\begin{array}{r}
75 \\
40\,\overline{\smash{\big)}\,3000} \\
\underline{280} \\
200
\end{array}
$$

so

$$
\frac{3}{40} = 75 \times \frac{1}{10^3} = .075.
$$

Thus the use of the division process to obtain the decimal representation of 3/40 can be fully justified by showing that it is equivalent to the more correct method of expressing 3/40 strictly as a decimal fraction, namely $\frac{3}{40} = \frac{3}{40} \times \frac{25}{25} = \frac{75}{1000} = .075$; *but no such argument can possibly work for the fraction 3/39 = 1/13.* Indeed, if it were possible to express 1/13 as a decimal fraction, say

$$
\frac{1}{13} = \frac{a}{10^n} \qquad \text{for some whole numbers } a \text{ and } n
$$

then the number *a* would satisfy

$$
10^n = 13 \times a
$$

so 13 would have to divide $10^n$; but this is impossible (Exercise 1). Thus *3/39 = 1/13 cannot be written as a decimal fraction.*

However, as we remarked in Chapter II.2, school mathematics is *naively constructive* in spirit, and pays little heed to proofs of existence or nonexistence, possibility or impossibility. In most cases we scarcely notice this *fundamental weakness* of decimal fractions. When confronted with the challenge of finding *the decimal representation of 3/39 = 1/13* we more likely than not apply the same procedure which worked so well for the fraction 3/40: that is, we *try* to divide *1.000 ... by 13*, making use of as many zeros after the decimal point as we seem to require!

```
        .0 7 6 9 2 3 0 . . .
13 | 1.0 0 0 0 0 0 0 . . .
     9 1
     ───
       9 0
       7 8
       ───
       1 2 0
       1 1 7
       ─────
           3 0
           2 6
           ───
             4 0
             3 9
             ───
              1 0 .
                . .
```

In the case of 3/40, this procedure could be justified only because we needed to make use of just *three* zeros, whence

*dividing 3 by 40*

was seen to be essentially the same as

*dividing 3000 by 40*, and then *multiplying the answer by* $\dfrac{1}{10^3}$ .

In the case of 3/39 = 1/13, we naively applied the same procedure as was used for the fraction 3/40, *even though it could not possibly be justified in the same way*; in the case of 3/39, the procedure simply generates an endless sequence of digits

.0 7 6 9 2 3 0 7 6 9 2 3 0 7 6 9 2 3 0 7 6 9 2 3 0 7 6 9 . . .

whose precise meaning is obscure. The procedure still *works* in the sense that it generates an answer (of a sort); but the kind of answer it generates is quite unlike the kind of answer we had come to expect—it simply cannot be interpreted as a decimal fraction $a/10^n$.

For many students this tension, between

*the familiar division procedure*

on the one hand, and

*the uncomprehended answer it generates*

on the other hand, is never resolved. Perhaps the most common indication of this unresolved tension arises when the student is confronted simultaneously with the following "facts":

(i) his/her natural gut-feeling that

$.\dot{9} = .999 \ldots$ *is just less than 1*;

(ii) the (uncomprehended) process of dividing 1 by 9, 2 by 9, 3 by 9, etc., which is observed to yield

$$\frac{1}{9} = .111 \ldots, \quad \frac{2}{9} = .222 \ldots, \quad \frac{3}{9} = .333 \ldots, \quad \text{etc.;}$$

(iii) the apparent consequence of (ii), that

$$1 = \frac{9}{9} = 9 \times \frac{1}{9} = .999 \ldots .^{[1]}$$

But however uncomprehended and unexpected the answer

$$\frac{1}{13} = .076923076923076923076 \ldots$$

may be, the simple arithmetical facts which it seems to suggest, namely

$$.07 = \frac{7}{100} < \frac{1}{13} < \frac{8}{100} = .08$$

$$.076 = \frac{76}{1000} < \frac{1}{13} < \frac{77}{1000} = .077$$

$$.0769 = \frac{769}{10000} < \frac{1}{13} < \frac{770}{10000} = .0770$$

are easily seen to be quite correct—an observation which suggests that this as yet *uncomprehended* string of digits contains more than a grain of truth, and that this truth might become *comprehensible* if we could only attribute some *clear mathematical meaning* to such *infinite decimals*.

Suppose we *try* to interpret this endless sequence of digits *in the spirit* of decimal fractions. Since

$$.0769 = 0 \cdot \frac{1}{10} + 7 \cdot \frac{1}{10^2} + 6 \cdot \frac{1}{10^3} + 9 \cdot \frac{1}{10^4},$$

it is natural to expect that

$$.07692307692307 \ldots = 0 \cdot \frac{1}{10} + 7 \cdot \frac{1}{10^2} + 6 \cdot \frac{1}{10^3} + 9 \cdot \frac{1}{10^4} + 2 \cdot \frac{1}{10^5}$$

$$+ 3 \cdot \frac{1}{10^6} + 0 \cdot \frac{1}{10^7} + 7 \cdot \frac{1}{10^8} + \ldots$$

But however harmless this expression may appear at first sight, *these can be no ordinary addition signs.*

---

[1] Since the traditional approaches of most sixth form and undergraduate courses deem it unnecessary deliberately to plan such disturbing experiences as this for their students—assuming either that they are unnecessary or unimportant—it is not at all uncommon for this particular experience to have to wait until middle age or later.

The addition of elementary arithmetic allows us, in the first instance, to add any *two* numbers *a, b*. By repeating this process twice we can evaluate the sum of any *three* numbers *a, b, c*, and we can do so in two distinct ways

$$(a + b) + c \quad \text{or} \quad a + (b + c):$$

thus we can either add *a* and *b* first, and then add *c* to the result; or we can add *b* and *c* first and then add the result to *a*. It is one of the basic, and most important, facts of elementary arithmetic that these two processes always yield the same end result; thus we can drop the brackets and write the sum of three numbers *a, b, c* simply as

$$a + b + c.$$

By repeating the basic addition process three times we can evaluate the sum of any *four* numbers *a, b, c, d*, and we can do this in five different ways (Exercise 2)—for example, here are three of them:

$$((a + b) + c) + d, \quad \text{or } (a + b) + (c + d), \quad \text{or } a + ((b + c) + d).$$

It turns out that all these processes yield the same end result, so we can drop the brackets and write the sum of four numbers *a, b, c, d* simply as

$$a + b + c + d.$$

A similar argument allows us

to evaluate the sum of any *five* numbers, $a_1, a_2, a_3, a_4, a_5$, and to write their sum simply as $a_1 + a_2 + a_3 + a_4 + a_5$ ; and

to evaluate the sum of any *six* numbers $a_1, a_2, a_3, a_4, a_5, a_6$, and to write their sum simply as $a_1 + a_2 + a_3 + a_4 + a_5 + a_6$.

And in general we are free

to evaluate the sum of any *N* numbers $a_1, a_2, \ldots, a_N$ (*where N is any finite number*), and to write their sum simply as $a_1 + a_2 + \ldots + a_N$.

But though this can be proved for arbitrarily long *finite* sums, it tells us nothing at all about either the meaning or the behaviour of *endless* sums like

$$0 \cdot \frac{1}{10} + 7 \cdot \frac{1}{10^2} + 6 \cdot \frac{1}{10^3} + 9 \cdot \frac{1}{10^4} + 2 \cdot \frac{1}{10^5}$$

$$+ 3 \cdot \frac{1}{10^6} + 0 \cdot \frac{1}{10^7} + 7 \cdot \frac{1}{10^8} + \cdots$$

Yet students are often encouraged to smudge over the fundamental distinction between *endless* and *long finite* sums. We must therefore start by underlining this distinction as forcibly as we can.

Given any finite whole number *N* (no matter how large), I can always *in principle*[2] count from 1 up to $N$: but *it is complete nonsense to suggest that I could ever count* all *the positive whole numbers.*[3]

In exactly the same way, though the familiar addition of elementary arithmetic allows us to evaluate the sum of $N$ numbers

$$a_1 + a_2 + a_3 + \cdots + a_N,$$

no matter how large $N$ may be, there is no justification whatsoever for assuming that this immediately allows us to evaluate *endless* sums such as

$$0 \cdot \frac{1}{10} + 7 \cdot \frac{1}{10^2} + 6 \cdot \frac{1}{10^3} + 9 \cdot \frac{1}{10^4} + 2 \cdot \frac{1}{10^5} + 3 \cdot \frac{1}{10^6}$$

$$+ 0 \cdot \frac{1}{10^7} + 7 \frac{1}{10^8} + 6 \cdot \frac{1}{10^9} + \cdots$$

or

$$1 + \frac{1}{2} + \frac{1}{3} + \frac{1}{4} + \frac{1}{5} + \frac{1}{6} + \frac{1}{7} + \frac{1}{8} + \cdots$$

or

$$\frac{1}{2} + \frac{1}{3} + \frac{1}{5} + \frac{1}{7} + \frac{1}{11} + \frac{1}{13} + \frac{1}{17} + \frac{1}{19} + \frac{1}{23} + \cdots$$

or

$$1\frac{1}{2} + \left(-1\frac{1}{3}\right) + 1\frac{1}{4} + \left(-1\frac{1}{5}\right) + 1\frac{1}{6} + \left(-1\frac{1}{7}\right) + 1\frac{1}{8} + \cdots$$

or

$$1\frac{1}{2} + \left(-1\frac{1}{4}\right) + 1\frac{1}{8} + \left(-1\frac{1}{16}\right) + 1\frac{1}{32} + \left(-1\frac{1}{64}\right) + \cdots .$$

And though we observed that the familiar addition process of elementary arithmetic allows us to insert any brackets we please in the sum

$$a_1 + a_2 + a_3 + \cdots + a_N$$

---

[2] Of course you must either ignore the fact that I might die before I finish, or else allow me to speed up the process of counting. Thus if $N = 10^{10}$ and I count one number per second, then I should require about 300 years; if one can imagine counting one number every micro-second, then one could finish in 3 hours.

[3] Suppose, I have counted as far as some very large number $N$; then I am no better off than when I began, since the task of counting *all* the numbers

$$N + 1, N + 2, N + 3, N + 4, N + 5, \ldots$$

is exactly the same as the original task of counting all the positive whole numbers:

$$1, \quad 2, \quad 3, \quad 4, \quad 5, \ldots$$

without affecting the end result, *there is no reason whatever to assume that the same will necessarily be true for endless sums* (even when we manage to explain just how such endless sums are to be understood): for example, one way of inserting brackets in the last example above produces the following

$$\left[1\frac{1}{2}+\left(-1\frac{1}{4}\right)\right]+\left[1\frac{1}{8}+\left(-1\frac{1}{16}\right)\right]$$

$$+\left[1\frac{1}{32}+\left(-1\frac{1}{64}\right)\right]+\ldots=\frac{1}{4}+\frac{1}{16}+\frac{1}{64}+\ldots,$$

whereas a slightly different set of brackets produces

$$1\frac{1}{2}+\left[\left(-1\frac{1}{4}\right)+1\frac{1}{8}\right]+\left[\left(-1\frac{1}{16}\right)+1\frac{1}{32}\right]$$

$$+\left[\left(-1\frac{1}{64}\right)+1\frac{1}{128}\right]+\ldots=\frac{3}{2}-\frac{1}{8}-\frac{1}{32}-\frac{1}{128}-\ldots,$$

and it is not at all clear that we should expect the two right hand sides to be equal: indeed, the only meaning we shall give to endless sums attributes the values $1/3$ to the first right hand side "$1/4 + 1/16 + 1/64 + \ldots$," and $4/3$ to the second "$3/2 - 1/8 - 1/32 - 1/128 - \ldots$," and *attributes no sum at all to the original expression*

$$1\frac{1}{2}+\left(-1\frac{1}{4}\right)+1\frac{1}{8}+\left(-1\frac{1}{16}\right)+1\frac{1}{32}+\left(-1\frac{1}{64}\right)+\ldots.$$

One of the elementary results in the study of endless sums shows that *whenever we succeed in attributing an ordinary numerical value to such an endless sum, we may insert any brackets we please without changing this value.* In contrast, the way in which the numerical value of such an endless sum is defined depends fundamentally on the *order* in which the terms appear: thus *if the order of the terms is changed, then the resulting endless sum may have a different value, or no value at all* (Exercise 8).

### Exercises

1. On page 71, it was asserted that, whatever value of $n$ we choose,

*13 cannot divide $10^n$.*

that is,

*$13 \times a = 10^n$ is impossible (if $a$ is a whole number).*

Suppose $a$ has units digit equal to $i$: that is,

$$a = 10 \times b + i \qquad (0 \le i \le 9).$$

Let $j$ be the units digit of $13 \times a$: that is,

$$13 \times a = 10 \times c + j \qquad (0 \le j \le 9)$$

For each value of $i$, find the corresponding value of $j$:

| units digit of $a$: $= i$ | 0 | 1 | 2 | 3 | 4 | 5 | 6 | 7 | 8 | 9 |
|---|---|---|---|---|---|---|---|---|---|---|
| units digit of $13 \times a$: $= j$ | | | | | | | | | | |

Use this to prove (by induction on $n$, or otherwise) that

$$13 \times a = 10^n \text{ is impossible.}$$

2. (i) Show that the sum of *four* numbers $a$, $b$, $c$, $d$ can be evaluated in just five different ways.

(ii) In how many different ways can the sum of five numbers $a$, $b$, $c$, $d$, $e$ be evaluated? Explain! (It is not too hard to prove the slightly curious fact that the number of different ways of inserting brackets in a sum of $n$ terms

$$a_1 + a_2 + a_3 + \cdots + a_n$$

is exactly the same as the number of different ways of inserting $n - 2$ diagonals in a convex $(n + 1)$-gon so as to dissect it into triangles. The number of such dissections of an $(n + 1)$-gon is discussed in Chapter 12 of *Mathematical Gems* by Ross Honsberger—published by the Mathematical Association of America (1973).)

3. Use the familiar division procedure to express the following fractions in decimal form:

$$\frac{1}{2}, \frac{1}{3}, \frac{1}{4}, \frac{1}{5}, \frac{1}{6}, \frac{1}{7}, \frac{1}{8}, \frac{1}{9}, \frac{1}{10}, \frac{1}{11}, \frac{1}{12}, \cdots$$

$$\frac{3}{2}, \frac{2}{3}, \frac{3}{4}, \frac{2}{5}, \frac{5}{6}, \frac{2}{7}, \frac{3}{8}, \frac{2}{9}, \frac{3}{10}, \frac{2}{11}, \frac{5}{12}, \cdots$$

4. For some whole numbers $n$, $1/n$ can be expressed as a proper decimal fraction (that is, the decimal representation of $1/n$ *terminates*); for other values of $n$, the familiar division procedure produces an infinite decimal representation of $1/n$. Give a complete description of those values of $n$, for which the decimal representation of $1/n$ terminates.

5. Use the familiar division procedure to express the following fractions in decimal form:

(i) $\dfrac{1}{9}, \dfrac{1}{99}, \dfrac{1}{999}, \dfrac{1}{9999}, \cdots$

(ii) $\dfrac{1}{1}, \dfrac{1}{11}, \dfrac{1}{111}, \dfrac{1}{1111}, \cdots$

6. Whole numbers can be expressed in terms of any base we choose; thus

$$twenty\ six = 2 \times 10^1 + 6 = 26_{base\ 10}$$

$$= 1 \times 5^2 + 0 \times 5 + 1 = 101_{base\ 5}$$

$$= 4 \times 6^1 + 1 = 41_{base\ 6}$$

$$= 1 \times 2^4 + 1 \times 2^3 + 0 \times 2^2 + 1 \times 2 + 0 = 11010_{base\ 2}$$

Express the following whole numbers in terms of *base 6*:
$7 = 1 \times 6 + 1 = \underline{11}_{base\ 6}$ ; $8 = \underline{\hspace{1cm}}_{base\ 6}$ ; $9 = \underline{\hspace{1cm}}_{base\ 6}$ ; $15 = \underline{\hspace{1cm}}_{base\ 6}$ ;
$16 = \underline{\hspace{1cm}}_{base\ 6}$ ; $35 = \underline{\hspace{1cm}}_{base\ 6}$ ; $37 = \underline{\hspace{1cm}}_{base\ 6}$ ; $100 = \underline{\hspace{1cm}}_{base\ 6}$ .

7. Fractions can also be expressed with respect to a variety of bases; thus

$$\frac{1}{4} = 25 \cdot \frac{1}{10^2} = \cdot 25_{base\ 10}$$

$$= 9 \cdot \frac{1}{6^2} \quad = (6+3) \cdot \frac{1}{6^2} = 1 \cdot \frac{1}{6} + 3 \cdot \frac{1}{6^2} = \cdot 13_{base\ 6}$$

$$= 1 \cdot \frac{1}{2^2} \quad = \cdot 01_{base\ 2}$$

In *base 10* the corresponding expression is called a *decimal*.
In *base 6* we shall call the corresponding expression a *heximal*.

(i) Which fraction has heximal .1?
   Which fraction has heximal .01?
   Which fraction has heximal .001?

(ii) Find the heximal representation[4] of the following ordinary fractions:

$$\frac{1}{2}, \frac{1}{3}, \frac{1}{4}, \frac{1}{5}, \frac{1}{6}, \frac{1}{7}, \frac{1}{8}, \frac{1}{9}, \frac{1}{10}, \frac{1}{11}, \frac{1}{12}, \ldots$$

(iii) Give a complete description of those values of $n$, for which the heximal representation of $1/n$ terminates.

(iv) Which ordinary fractions have heximal representation

$$\dot{.1} = .111 \ldots {}_{\cdot base\ 6}, \quad \dot{.0}\dot{1} = .010101 \ldots {}_{\cdot base\ 6}, \quad \dot{.00}\dot{1}_{base\ 6}, \ldots$$

---

[4] The first three can be done purely in terms of fractions: for example

$$\frac{1}{4} = 9 \times \frac{1}{6^2} = (6+3) \cdot \frac{1}{6^2} = 1 \cdot \frac{1}{6} + 3 \cdot \frac{1}{6^2} = .13_{base\ 6}$$

For those which cannot be handled in this way we can always express numerator and denominator in *base 6* and carry out a *base 6 long division*: for example

```
        . 1 1 . . .
      _____
  5 | 1 . 0 0 0 0 0 . . .
      5
      _____
      1 0
        5
      _____
```

(v)  Which ordinary fractions have heximal representation:

$$.\dot{5} = .555\ldots_{\cdot base\ 6}, \qquad .0\dot{5}, \quad .00\dot{5}, \ldots$$

8. (i) Let $s_n = \sum_{r=1}^{n} [(-1)^{r+1}/r]$ be the sum of the first $n$ terms of the endless sum

$$1 - \frac{1}{2} + \frac{1}{3} - \frac{1}{4} + \frac{1}{5} - \frac{1}{6} + \frac{1}{7} - \ldots$$

(a)  Show that

$$s_2 < s_4 < s_6 < \ldots < s_5 < s_3 < s_1.$$

(b)  Prove that, for each $n, m \geq 1$,

$$s_{2n} < s_{2n+2} < s_{2m+1} < s_{2m-1}$$

and that

$$s_{2n+1} - s_{2n} = \frac{1}{2n+1}.$$

(c)  Deduce that, for $n \geq 2$,

$$\frac{1}{2} \leq s_n \leq \frac{5}{6}.$$

(ii) Let $t_n$ be the sum of the first $n$ terms of the endless sum

$$1 - \frac{1}{2} - \frac{1}{4} + \frac{1}{3} - \frac{1}{6} - \frac{1}{8} + \frac{1}{5} - \frac{1}{10} - \frac{1}{12} + \frac{1}{7} - \frac{1}{14} - \ldots$$

(which looks like the endless sum of part (i) written in a slightly different order).

(a)  Show that

$$t_{3n-1} = \tfrac{1}{2}s_{2n-1}, \qquad t_{3n} = \tfrac{1}{2}s_{2n}, \qquad t_{3n+1} = \tfrac{1}{2}s_{2n} + \frac{1}{2n-1}.$$

(b)  Deduce that, for $n \geq 3$,

$$t_n \leq \frac{5}{12} < \frac{1}{2} \leq s_n$$

(iii) Let $t'_n$ be the sum of the first $n$ terms of the endless sum

$$1 + \frac{1}{3} - \frac{1}{2} + \frac{1}{5} + \frac{1}{7} - \frac{1}{4} + \frac{1}{9} + \frac{1}{11} - \frac{1}{13} + \ldots$$

(which looks like the endless sum of part (i) written in a slightly different order).

(a)  Show that

$$t'_1 < t'_4 < t'_7 < \ldots < t'_8 < t'_5 < t'_2$$

$$\text{and} \quad t'_3 < t'_6 < t'_9 < \ldots < t'_8 < t'_5 < t'_2.$$

(b) Prove that, for each $n, m \geq 1$,

$$t'_{3n-2} < t'_{3n+1} < t'_{3m+2} < t'_{3m-1}$$

$$\text{and} \quad t'_{3n} < t'_{3n+3} < t'_{3m+2} < t'_{3m-1}$$

and that

$$t'_{3n-1} - t'_{3n-2} = \frac{1}{4n-1} \quad \text{and} \quad t'_{3n-1} - t'_{3n} = \frac{1}{2n}.$$

(c) Deduce that, for $n \geq 2$,

$$s_n \leq \frac{5}{6} < \frac{389}{420} \leq t'_n.$$

CHAPTER II.7

# Infinite Decimals (Part 2)

In Chapter II.6 we stressed the tension between

the *familiar* division process

used to generate the decimal representation of a fraction like 3/40, and

the (as yet) *meaningless* string of digits

which results when we apply this same familiar division process to a fraction like $3/39 = 1/13$, namely

$$\frac{1}{13} = .07692307692307692307\ldots$$

We also observed that this endless sequence of digits appeared nevertheless to represent 1/13 in a way which warranted further study. As a first tentative step we rewrote this *infinite decimal* as nearly as possible in the spirit of ordinary decimal fractions. Now, strictly speaking, $1/8 = .125$ corresponds to the decimal fraction 125/1000, but for *infinite* decimals this interpretation is a complete non-starter; thus we optimistically tried to generalise the alternative interpretation $1/8 = 1 \cdot (1/10) + 2 \cdot (1/10^2) + 5 \cdot (1/10^3)$, and so produced the *endless sum*

$$\frac{1}{13} = 0 \cdot \frac{1}{10} + 7 \cdot \frac{1}{10^2} + 6 \cdot \frac{1}{10^3} + 9 \cdot \frac{1}{10^4}$$

$$+ 2 \cdot \frac{1}{10^5} + 3 \cdot \frac{1}{10^6} + 0 \cdot \frac{1}{10^7} + 7 \cdot \frac{1}{10^8} + \ldots$$

whose precise meaning remains obscure, since ordinary addition only allows us to add together finitely many terms. This we shall now try to clarify. In our attempt to do so we shall confront the simplest example of that fundamental extension procedure by which we transcend the *finite processes* of elementary mathematics (arithmetic, algebra, and school geometry) and succeed in capturing those *infinite processes* which lie at the root of the mathematics of continuous magnitudes. And it is precisely such infinite processes which hold the key to any acceptable foundation for the calculus.

In fumbling to make some sense of the endless sum above we cannot pretend to make strict deductions, for we have not as yet attributed any strictly mathematical meaning to this curious expression. Nevertheless, it is natural to exploit the intrinsic recurring pattern, and to bracket the terms accordingly, in the hope that the resulting expression may be easier to manage: thus it is tempting to rewrite our endless sum in the form

$$\left(0 \cdot \frac{1}{10} + 7 \cdot \frac{1}{10^2} + 6 \cdot \frac{1}{10^3} + 9 \cdot \frac{1}{10^4} + 2 \cdot \frac{1}{10^5} + 3 \cdot \frac{1}{10^6}\right)$$

$$+ \left(0 \cdot \frac{1}{10^7} + 7 \cdot \frac{1}{10^8} + 6 \cdot \frac{1}{10^9} + 9 \cdot \frac{1}{10^{10}} + 2 \cdot \frac{1}{10^{11}} + 3 \cdot \frac{1}{10^{12}}\right) + \cdots$$

$$= \frac{76923}{10^6} + \frac{76923}{10^{12}} + \frac{76923}{10^{18}} + \cdots .$$

Now for finitely many terms $a_1, a_2, \ldots, a_N$ we know that the distributive law allows us to *take out common factors*:

$$xa_1 + xa_2 + xa_3 + \ldots + xa_N = x(a_1 + a_2 + a_3 + \ldots + a_N).$$

Once again, though we cannot possibly justify the move at this stage, in trying to reduce our present task to the simplest possible form it is natural to rewrite the above endless sum in the form

$$\frac{76923}{10^6} \left(1 + \frac{1}{10^6} + \frac{1}{10^{12}} + \frac{1}{10^{18}} + \cdots\right)$$

It would therefore appear that it might be sensible to start by trying to make sense of the expression in brackets, namely

$$1 + \frac{1}{10^6} + \frac{1}{10^{12}} + \frac{1}{10^{18}} + \cdots$$

(which should turn out to correspond to the infinite decimal

$$1.00000100000100000100 \ldots :$$

see Exercise 5(i) in Chapter II.6).

This problem is a particular example of a whole family of problems of exactly the same kind: namely

*For what values of x can we give some precise mathematical meaning to the endless sum*

$$1 + x + x^2 + x^3 + x^4 + \ldots ?$$

In our particular case $x$ is $1/10^6$, but it is no more difficult, and perhaps more instructive, to consider all such examples at the same time. The reader who thinks she/he has "done GP's" is nevertheless advised to resist any temptation to yawn, or to recite the magic formula $a/(1 - r)$. Mathematicians as great as John Wallis (1616–1703) and Leibniz appealed to this magic formula to justify writing such curiosities as

$$1 + 2 + 4 + 8 + 16 + \ldots = \frac{1}{1-2} = -1$$

and

$$1 - 1 + 1 - 1 + 1 - 1 + \ldots = \frac{1}{1-(-1)} = \frac{1}{2}$$

and we would all do well to consider carefully the conditions under which endless sums of the form

$$1 + x + x^2 + x^3 + x^4 + \ldots$$

can be given a precise mathematical meaning.

As a first step observe that *the endless sum*

$$1 + x + x^2 + x^3 + x^4 + \ldots$$

can be reconstructed from *the endless sequence of finite sums*

$$1, \quad 1 + x, \quad 1 + x + x^2, \quad 1 + x + x^2 + x^3, \quad 1 + x + x^2 + x^3 + x^4, \ldots .$$

And given a number $x$, each of these *finite* sums can be evaluated by the ordinary operations of elementary arithmetic; so this endless sequence of finite sums is really just an *endless sequence of ordinary numbers*.

EXAMPLE A. If $x = 1/10$, then the endless sum

$$1 + \frac{1}{10} + \frac{1}{10^2} + \frac{1}{10^3} + \frac{1}{10^4} + \ldots$$

can be reconstructed from the endless sequence of finite sums

$$1, \quad 1 + \frac{1}{10}, \quad 1 + \frac{1}{10} + \frac{1}{10^2}, \quad 1 + \frac{1}{10} + \frac{1}{10^2} + \frac{1}{10^3},$$

$$1 + \frac{1}{10} + \frac{1}{10^2} + \frac{1}{10^3} + \frac{1}{10^4}, \ldots$$

each of which can be evaluated—so this endless sequence of finite sums is really just the endless sequence

$$1, \quad \frac{11}{10} = 1.1, \quad \frac{111}{100} = 1.11, \quad \frac{1111}{1000} = 1.111, \quad \frac{11111}{10000} = 1.1111, \ldots$$

of ordinary numbers.

EXAMPLE B. If $x = -1$, then the endless sum

$$1 + (-1) + (-1)^2 + (-1)^3 + (-1)^4 + \ldots$$

can be reconstructed from the endless sequence of finite sums

$$1, \quad 1-1, \quad 1-1+1, \quad 1-1+1-1, \quad 1-1+1-1+1, \ldots$$

each of which can be evaluated—so this endless sequence of finite sums is really the endless sequence

$$1, \quad 0, \quad 1, \quad 0, \quad 1, \quad \ldots$$

of ordinary numbers.

EXAMPLE C. If $x = 2$, then the endless sum

$$1 + 2 + 2^2 + 2^3 + 2^4 + \ldots$$

can be reconstructed from the endless sequence of finite sums

$$1, \quad 1+2, \quad 1+2+2^2, \quad 1+2+2^2+2^3, \quad 1+2+2^2+2^3+2^4, \ldots$$

each of which can be evaluated—so this endless sequence of finite sums is really the endless sequence

$$1, \quad 3, \quad 7, \quad 15, \quad 31, \ldots$$

of ordinary numbers.

EXAMPLE D. If $x = -1/2$, then the endless sum

$$1 + \left(-\frac{1}{2}\right) + \left(-\frac{1}{2}\right)^2 + \left(-\frac{1}{2}\right)^3 + \left(-\frac{1}{2}\right)^4 + \ldots$$

can be reconstructed from the endless sequence of finite sums

$$1, \quad 1-\frac{1}{2}, \quad 1-\frac{1}{2}+\frac{1}{4}, \quad 1-\frac{1}{2}+\frac{1}{4}-\frac{1}{8}, \quad 1-\frac{1}{2}+\frac{1}{4}-\frac{1}{8}+\frac{1}{16}, \ldots$$

each of which can be evaluated—so this endless sequence of finite sums is really just the endless sequence

$$1, \quad \frac{1}{2}, \quad \frac{3}{4}, \quad \frac{5}{8}, \quad \frac{11}{16}, \ldots$$

of ordinary numbers.

Thus our original task of *attaching a numerical value to an endless sum* of the form

$$1 + x + x^2 + x^3 + x^4 + \ldots$$

(where $x$ denotes some given number), has now been transformed into the task of *attaching a numerical value to the endless sequence of ordinary numbers*

$$1, \quad 1+x, \quad 1+x+x^2, \quad 1+x+x^2+x^3, \quad 1+x+x^2+x^3+x^4, \ldots.$$

And though this change of viewpoint may not look very encouraging, it has in fact brought us very close to the solution of our original problem! For by elementary algebra we have

$$(1 + x)(1 - x) = 1 - x^2,$$

$$(1 + x + x^2)(1 - x) = 1 - x^3,$$

$$(1 + x + x^2 + x^3)(1 - x) = 1 - x^4,$$

and, in general,

$$(1 + x + x^2 + x^3 + \ldots + x^n)(1 - x) = 1 - x^{n+1}.$$

Thus, *as long as x is not equal to 1,*[1] we have

$$1 + x + x^2 + x^3 + \ldots + x^n = \frac{1 - x^{n+1}}{1 - x} = \frac{1}{1 - x} + \frac{x^{n+1}}{x - 1},$$

so that, provided $x \neq 1$, we may write our endless sequence of ordinary numbers

$$1, \quad 1 + x, \quad 1 + x + x^2, \quad 1 + x + x^2 + x^3, \ldots$$

in the alternative form

$$\frac{1}{1 - x} + \frac{x}{x - 1}, \quad \frac{1}{1 - x} + \frac{x^2}{x - 1}, \quad \frac{1}{1 - x} + \frac{x^3}{x - 1}, \quad \frac{1}{1 - x} + \frac{x^4}{x - 1}, \ldots.$$

Thus for any given number $x$ (except for $x = 1$), each number in this sequence is equal to $1/(1 - x)$ plus something that we shall simply call the *error term* $x^n/(x - 1)$.

If $x = 0$, then each "error term" is equal to 0, and every term of the endless sequence of finite sums is equal to $1/(1 - 0) = 1$; we therefore, not surprisingly, assign the numerical value 1 to this sequence, and hence also to the endless sum

$$1 + 0 + 0^2 + 0^3 + 0^4 + \ldots.$$

If $x$ is a given number strictly between $-1$ and $+1$, then the denominator $x - 1$ of the "error term" $x^n/(x - 1)$ is fixed, whereas the numerator $x^n$ rapidly approaches zero as $n$ increases (Exercises 1, 2, 4): thus the terms of our endless sequence

$$\frac{1}{1 - x} + \frac{x}{x - 1}, \quad \frac{1}{1 - x} + \frac{x^2}{x - 1}, \quad \frac{1}{1 - x} + \frac{x^3}{x - 1}, \quad \frac{1}{1 - x} + \frac{x^4}{x - 1}, \ldots$$

differ from $1/(1 - x)$ by an amount which decreases rapidly towards zero;

---

[1] *If $x = 1$, then we have the endless sum*

$$1 + 1 + 1 + 1 + 1 + \ldots$$

to which we would scarcely expect to assign a numerical value!

and though, for $x \neq 0$, the "error term" never disappears altogether, it is natural to interpret this sequence as *tending towards*

$$\frac{1}{1-x}$$

in some genuine sense; thus we assign this numerical value to the endless sequence, and hence also to the endless sum

$$1 + x + x^2 + x^3 + x^4 + \dots .^2$$

*If $x = -1$*, then the "error term" alternates between $(-1)^{2n-1}/(-1-1)$ $= +1/2$ and $(-1)^{2n}/(-1-1) = -1/2$, so the sequence *oscillates*

$$1, \quad 0, \quad 1, \quad 0, \quad 1, \quad 0, \quad 1, \dots .$$

In such a case we make no attempt to assign a numerical value either to the endless sequence, or to the endless sum

$$1 + (-1) + (-1)^2 + (-1)^3 + (-1)^4 + \dots$$

*If $x$ is strictly greater than $+1$*, then the denominator $x - 1$ of the "error term" $x^n/(x-1)$ is fixed and positive, whereas the numerator $x^n$ grows rapidly as $n$ increases and eventually exceeds any given positive whole number $N$ (Exercises 1, 2, 3); the terms of our endless sequence

$$\frac{1}{1-x} + \frac{x}{x-1}, \quad \frac{1}{1-x} + \frac{x^2}{x-1}, \quad \frac{1}{1-x} + \frac{x^3}{x-1}, \quad \frac{1}{1-x} + \frac{x^4}{x-1}, \dots$$

grow in a similar way, In such a case we make no attempt to assign a numerical value either to the endless sequence, or to the endless sum

$$1 + x + x^2 + x^3 + x^4 \dots .$$

*If $x$ is strictly less than $-1$*, then the denominator $x - 1$ of the "error term" $x^n/(x-1)$ is fixed and negative, while the numerator $x^n$ alternates in sign, but grows rapidly *in absolute value* as $n$ increases. Moreover, it does so in such a way that, given any positive whole number $N$, $x^n$ eventually surpasses both $N$ and $-N$ (Exercise 3). The terms of our endless sequence

$$\frac{1}{1-x} + \frac{x}{x-1}, \quad \frac{1}{1-x} + \frac{x^2}{x-1}, \quad \frac{1}{1-x} + \frac{x^3}{x-1}, \quad \frac{1}{1-x} + \frac{x^4}{x-1}, \dots$$

oscillate and grow in a similar way. In such a case we make no attempt to assign a numerical value either to the endless sequence, or to the endless sum

$$1 + x + x^2 + x^3 + x^4 + \cdots$$

*Thus we interpret the endless sum*

$$1 + x + x^2 + x^3 + x^4 + \cdots$$

---

[2] Note that when $x = 0$, this agrees with the conclusion of the previous paragraph.

*as a kind of endpoint towards which the endless sequence of finite sums*

$$1, \quad 1 + x, \quad 1 + x + x^2, \quad 1 + x + x^2 + x^3, \dots,$$

*is heading.* For $x \neq 1$, this sequence can be written in the alternative form

$$\frac{1}{1-x} + \frac{x}{x-1}, \quad \frac{1}{1-x} + \frac{x^2}{x-1}, \quad \frac{1}{1-x} + \frac{x^3}{x-1}, \quad \frac{1}{1-x} + \frac{x^4}{x-1}, \dots.$$

*If $x$ lies strictly between $-1$ and $1$, then this sequence is clearly heading for*

$$\frac{1}{1-x}$$

*and we attach precisely this numerical value to the endless sum*

$$1 + x + x^2 + x^3 + x^4 + \dots.$$

*For no other values of $x$ do we attempt to assign a numerical value to the endless sum $1 + x + x^2 + x^3 + \cdots$.*

We can now see that in Example A above it might have been more enlightening to transform the endless sequence

$$1, \quad 1 + \frac{1}{10}, \quad 1 + \frac{1}{10} + \frac{1}{10^2}, \quad 1 + \frac{1}{10} + \frac{1}{10^2} + \frac{1}{10^3}, \dots$$

into the alternative form

$$\frac{1}{1 - (1/10)} + \frac{(1/10)}{(1/10) - 1}, \quad \frac{1}{1 - (1/10)} + \frac{(1/10)^2}{(1/10) - 1},$$

$$\frac{1}{1 - (1/10)} + \frac{(1/10)^3}{(1/10) - 1}, \quad \frac{1}{1 - (1/10)} + \frac{(1/10)^4}{(1/10) - 1}, \dots$$

that is,

$$\frac{10}{9} - \frac{1}{9}, \quad \frac{10}{9} - \frac{1}{90}, \quad \frac{10}{9} - \frac{1}{900}, \quad \frac{10}{9} - \frac{1}{9000}, \dots$$

whence it is clear that the *finite* sums $1 + x + x^2 + \dots + x^n$ differ from $10/9$ by an amount $(= 1/(9 \times 10^n))$ which rapidly approaches zero as $n$ increases. We thus assign the numerical value $10/9$ to the endless sum

$$1 + \frac{1}{10} + \frac{1}{10^2} + \frac{1}{10^3} + \frac{1}{10^4} + \cdots$$

(which agrees with the naive feeling that the sequence 1, 1.1, 1.11, 1.111, 1.111, ... on page 83 is approaching the value $1.\dot{1}$—see Exercise 5(i) in Chapter II.6).

In Example B the endlessly oscillating behavior of the endless sequence

$$1, \quad 0, \quad 1, \quad 0, \quad 1, \quad 0, \quad 1, \quad 0, \dots$$

of finite sums is plain enough.

In Example C, although the unrestricted growth of the finite sums is more or less obvious, the alternative form

$$\frac{1}{1-2}+\frac{2}{2-1}, \quad \frac{1}{1-2}+\frac{2^2}{2-1}, \quad \frac{1}{1-2}+\frac{2^3}{2-1},$$

$$\frac{1}{1-2}+\frac{2^4}{2-1}, \quad \frac{1}{1-2}+\frac{2^5}{2-1}, \dots$$

or

$$-1+2, \quad -1+2^2, \quad -1+2^3, \quad -1+2^4, \quad -1+2^5, \dots$$

exhibits much more clearly the pattern underlying

$$1, \quad 3, \quad 7, \quad 15, \quad 31, \dots .$$

In Example D the endless sequence of finite sums

$$1, \quad \frac{1}{2}, \quad \frac{3}{4}, \quad \frac{5}{8}, \quad \frac{11}{16}, \dots$$

is nowhere near as suggestive as the alternative form

$$\frac{1}{1+(1/2)}+\frac{(-1/2)}{-(1/2)-1}, \quad \frac{1}{1+(1/2)}+\frac{(-1/2)^2}{-(1/2)-1},$$

$$\frac{1}{1+(1/2)}+\frac{(-1/2)^3}{-(1/2)-1}, \quad \frac{1}{1+(1/2)}+\frac{(-1/2)^4}{-(1/2)-1}, \dots$$

that is,

$$\frac{2}{3}+\frac{1}{3}, \quad \frac{2}{3}-\frac{1}{3\cdot 2}, \quad \frac{2}{3}+\frac{1}{3\cdot 2^2}, \quad \frac{2}{3}-\frac{1}{3\cdot 2^3}, \quad \frac{2}{3}+\frac{1}{3\cdot 2^4}, \dots .$$

Returning to the endless sum with which we began, namely

$$1+\frac{1}{10^6}+\frac{1}{10^{12}}+\frac{1}{10^{18}}+\frac{1}{10^{24}}+\dots,$$

we see that this corresponds to the endless sequence of finite sums

$$1, \quad 1+\frac{1}{10^6}, \quad 1+\frac{1}{10^6}+\frac{1}{10^{12}}, \quad 1+\frac{1}{10^6}+\frac{1}{10^{12}}+\frac{1}{10^{18}}, \dots$$

whose terms may be written in the alternative form

$$\frac{1}{1-(1/10^6)}+\frac{(1/10^6)}{(1/10^6)-1}, \quad \frac{1}{1-(1/10^6)}+\frac{(1/10^6)^2}{(1/10^6)-1},$$

$$\frac{1}{1-(1/10^6)}+\frac{(1/10^6)^3}{(1/10^6)-1}, \quad \frac{1}{1-(1/10^6)}+\frac{(1/10^6)^4}{(1/10^6)-1}, \dots .$$

that is,

$$\frac{10^6}{10^6 - 1} - \frac{1}{10^6 - 1}, \quad \frac{10^6}{10^6 - 1} - \frac{1}{(10^6 - 1) \cdot 10^6},$$

$$\frac{10^6}{10^6 - 1} - \frac{1}{(10^6 - 1) \cdot (10^6)^2}, \quad \frac{10^6}{10^6 - 1} - \frac{1}{(10^6 - 1) \cdot (10^6)^3}, \dots.$$

The endless sequence of *finite* sums is thus clearly heading for

$$\frac{10^6}{10^6 - 1}$$

(though it never quite arrives!), so we assign this numerical value to the *endless* sum

$$1 + \frac{1}{10^6} + \frac{1}{10^{12}} + \frac{1}{10^{18}} + \frac{1}{10^{24}} + \dots.$$

All this suggests that the endless sum

$$0 \cdot \frac{1}{10} + 7 \cdot \frac{1}{10^2} + 6 \cdot \frac{1}{10^3} + 9 \cdot \frac{1}{10^4} + 2 \cdot \frac{1}{10^5}$$

$$+ 3 \cdot \frac{1}{10^6} + 0 \cdot \frac{1}{10^7} + 7 \cdot \frac{1}{10^8} + \dots,$$

which we optimistically rewrote first in the form

$$\frac{76923}{10^6} + \frac{76923}{10^{12}} + \frac{76923}{10^{18}} + \frac{76923}{10^{24}} + \dots$$

and then in the form

$$\frac{76923}{10^6} \left(1 + \frac{1}{10^6} + \frac{1}{10^{12}} + \frac{1}{10^{18}} + \dots \right),$$

should be assigned the value

$$\frac{76923}{10^6} \times \frac{10^6}{10^6 - 1}.$$

But then the fact that

$$10^6 - 1 = 76923 \times 13$$

allows us to conclude that

*if* the expression

$$.076923076923076923076 \dots = .\overline{076923}$$

is first interpreted as an endless sum

$$0 \cdot \frac{1}{10} + 7 \cdot \frac{1}{10^2} + 6 \cdot \frac{1}{10^3} + 9 \cdot \frac{1}{10^4} + 2 \cdot \frac{1}{10^5}$$

$$+ 3 \cdot \frac{1}{10^6} + 0 \cdot \frac{1}{10^7} + 7 \cdot \frac{1}{10^8} + \dots,$$

*if* this endless sum is then rewritten in the form

$$\frac{76923}{10^6}\left(1 + \frac{1}{10^6} + \frac{1}{10^{12}} + \frac{1}{10^{18}} + \dots\right), \text{ and}$$

*if*, finally, the endless sum in brackets is interpreted precisely as was done above (that is, as the "endpoint" of the corresponding sequence of finite sums),
*then* we can sensibly write

$$.0\dot{7}692\dot{3} = \frac{1}{13}.$$

In closing this chapter we should remark that we took the as yet unjustified step from

$$\frac{76923}{10^6} + \frac{76923}{10^{12}} + \frac{76923}{10^{18}} + \frac{76923}{10^{24}} + \dots$$

to

$$\frac{76923}{10^6}\left(1 + \frac{1}{10^6} + \frac{1}{10^{12}} + \frac{1}{10^{18}} + \dots\right)$$

only to get rid of the awkward looking factor 76923, and to bring out the essential form of the original endless sum. But *we shall now show that*, though this appeared at the time to simplify our task, *it was not strictly necessary.*

Consider any endless sum of the form

$$a + ax + ax^2 + ax^3 + ax^4 + \dots$$

(in the particular example above $a = 76923/10^6$ and $x = 1/10^6$). This endless sum can clearly be reconstructed from the endless sequence of finite sums

$$a, \quad a + ax, \quad a + ax + ax^2, \quad a + ax + ax^2 + ax^3, \dots .$$

Since these are *finite* sums, the laws of ordinary arithmetic allow us to express the terms of this endless sequence in the alternative form

$$a, \quad a(1 + x), \quad a(1 + x + x^2), \quad a(1 + x + x^2 + x^3), \dots .$$

We can now proceed as before, and, as long as $x$ is not equal to $1^3$, we can use the fact that

$$1 + x + x + \dots + x^n = \frac{1}{1 - x} + \frac{x^{n+1}}{x - 1}$$

---

[3] If $a$ is not equal to 0 and $x = 1$, then we have the endless sum

$$a + a + a + a + \dots$$

to which we do not attempt to assign a numerical value.

to write this endless sequence of finite sums in the form

$$a\left(\frac{1}{1-x}+\frac{x}{x-1}\right),\quad a\left(\frac{1}{1-x}+\frac{x^2}{x-1}\right),$$

$$a\left(\frac{1}{1-x}+\frac{x^3}{x-1}\right),\quad a\left(\frac{1}{1-x}+\frac{x^4}{x-1}\right),\dots .$$

Thus each finite sum is equal to $a/(1-x)$ plus what we shall simply call the *error term* $ax^n/(x-1)$. We can now argue more or less as before, keeping half an eye free to take care of the possible values of the number $a$.

If $a = 0$, then, no matter what the value of $x$ may be, every finite sum is equal to 0; we therefore assign the numerical value 0 both to the endless sequence of finite sums, and to the endless sum

$$0+0\cdot x+0\cdot x^2+0\cdot x^3+0\cdot x^4+\dots .$$

*From now on we shall assume that a is some fixed number not equal to 0.*

If $x = 0$, then each of the error terms $ax^n/(x-1)$ is equal to 0, and every term of the endless sequence of finite sums is equal to $a/(1-0) = a$; we therefore assign the numerical value $a$ to the endless sequence of finite sums, and so also to the endless sum

$$a+a\cdot 0+a\cdot 0^2+a\cdot 0^3+a\cdot 0^4+\dots .$$

*If x is a given number strictly between* $-1$ *and* $+1$, then the denominator $x-1$ of the error term $ax^n/(x-1)$ is fixed, while the numerator $a\cdot x^n$ rapidly approaches zero as $n$ increases (Exercise 4): thus the terms of our endless sequence of finite sums

$$\frac{a}{1-x}+\frac{ax}{x-1},\quad \frac{a}{1-x}+\frac{ax^2}{x-1},\quad \frac{a}{1-x}+\frac{ax^3}{x-1},\quad \frac{a}{1-x}+\frac{ax^4}{x-1},\dots$$

differ from $a/(1-x)$ by an amount which decreases rapidly towards zero; and though, for $x \neq 0$, the error term never disappears altogether, it is clear that this endless sequence is *heading towards*

$$\frac{a}{1-x}.$$

We therefore assign this numerical value to the endless sequence of finite sums, and hence also to the endless sum

$$a+ax+ax^2+ax^3+ax^4+\dots .$$

If $x = -1$, then the endless sequence of finite sums *oscillates*

$$a,\quad 0,\quad a,\quad 0,\quad a,\quad 0,\dots$$

and we make no attempt to assign a numerical value either to the endless sequence, or to the endless sum

$$a+a(-1)+a(-1)^2+a(-1)^3+a(-1)^4+\dots .$$

If $x$ is either strictly greater than 1, or strictly less than $-1$, then, as before, the error term $ax^n/(x-1)$ grows in a totally unrestrained way. In this case we make no attempt to assign a numerical value either to the endless sequence of finite sums, or to the endless sum

$$a + ax + ax^2 + ax^3 + ax^4 + \dots .$$

Thus, if $a = 0$, then, no matter what the value of $x$ may be, we assign the numerical value 0 to the endless sum

$$0 + 0 \cdot x + 0 \cdot x^2 + 0 \cdot x^3 + 0 \cdot x^4 + \dots ;$$

if $a \neq 0$, then we assign a numerical value to the endless sum

$$a + ax + ax^2 + ax^3 + ax^4 + \dots$$

only for values of $x$ strictly between $-1$ and 1, and the value we assign in this case is $a/(1-x)$.

EXERCISES

1. Work out the following (using logarithms or a calculator):

(i) $.5^4 =$___,  $.5^7 =$___,  $.5^{10} =$___,  $.5^{13} =$___,  $.5^{16} =$___;

$.9^4 =$___,  $.9^7 =$___,  $.9^{10} =$___,  $.9^{13} =$___,  $.9^{16} =$___;

$.99^4 =$___,  $.99^7 =$___,  $.99^{10} =$___,  $.99^{13} =$___,  $.99^{16} =$___;

(ii) $1.5^4 =$___,  $1.5^7 =$___,  $1.5^{10} =$___,  $1.5^{13} =$___,  $1.5^{16} =$___;

$1.1^4 =$___,  $1.1^7 =$___,  $1.1^{10} =$___,  $1.1^{13} =$___,  $1.1^{16} =$___;

$1.01^4 =$___,  $1.01^7 =$___,  $1.01^{10} =$___,  $1.01^{13} =$___,  $1.01^{16} =$___.

2. Procure two sheets of graph paper. On each sheet choose the origin at the centre of the page and draw the y-axis up the length of the page. Mark the x-axis from $-3/2$ to $3/2$ on each sheet; use the same scale for the y-axis.

(i) On the first sheet plot the following graphs *as accurately as you can* (use a fine point, and a different colour for each curve):

$$y = x^2, \quad y = x^4, \quad y = x^6, \quad y = x^8,$$
$$y = x^{10}, \quad y = x^{12}, \quad y = x^{14}, \quad y = x^{16}.$$

(ii) On the second sheet plot the following graphs *as accurately as you can* (use a fine point and a different colour for each curve):

$$y = x, \quad y = x^3, \quad y = x^5, \quad y = x^7,$$
$$y = x^9, \quad y = x^{11}, \quad y = x^{13}, \quad y = x^{15}.$$

3. Exercises 1(ii) and 2 suggest that the following is true: *For any given value of $x > 1$, the powers $x^n$ grow rapidly as n increases. Moreover, for any given $x > 1$, the powers $x^n$ grow in such a way that they eventually exceed any positive whole number N you care to choose.*

You should now *prove* this is true. [Hint: We are given some number $x > 1$. If we put

$$h = x - 1$$

then we can be sure that $h$ is positive.
Expand

$$x^n = (1 + h)^n$$

using the binomial theorem, and show that

$$x^n \geq 1 + nh.$$

Deduce that

if we choose $n > (99/h)$,      then $x^n > 100$;

if we choose $n > (999/h)$,      then $x^n > 1000$.

In general, suppose that $N$ is any given positive whole number, however large. Show that

$$\text{if we choose } n > (N - 1)/h, \text{ then } x^n > N.]$$

4. Exercises 1(i) and 2 suggest that the following is true: *For any given positive value of $x < 1$, the powers $x^n$ rapidly approach zero as $n$ increases. Moreover, for any given $x$, the powers $x^n$ shrink in such a way that they eventually get smaller than $1/N$, where $N$ may be any positive whole number you care to choose.*
   You should now *prove* this is true. [Hint: The obvious analogue of the method for $x > 1$ does not work. Instead we use the fact that if $0 < x < 1$, then $1/x > 1$.
If we put

$$y = \frac{1}{x}$$

then $y > 1$.
Given any positive whole number $N$, then (by Exercise 3) we may choose $n$ such that

$$y^n > N$$

whence

$$x^n = \frac{1}{y^n} < \frac{1}{N} .]$$

5. Perhaps the most common way of converting an infinite decimal like .111 ... to a fraction goes as follows:
   "Given $x = .111 \ldots$
Multiply by 10 to get

$$10x = 1.111 \ldots$$

(after all, we all know that multiplying by 10 shifts the decimal point one place to the right).

Now subtract to get

$$9x = 10x - x = 1.111 \ldots -0.111 \ldots = 1$$

(which seems too obvious for words).
    Hence

$$x = 1/9.\text{"}$$

This argument has one minor and one major weakness (though the conclusion happens to be correct).
    The whole argument ignores the fact that the only meaning one can give to infinite decimals is in terms of *endless sums*: thus

$$.111 \ldots = 1 \cdot \frac{1}{10} + 1 \cdot \frac{1}{10^2} + 1 \cdot \frac{1}{10^3} + \ldots .$$

Instead the argument simply assumes that .111 ... behaves like an ordinary finite decimal. The fact is that

"multiplying by 10 shifts the decimal point one place to the right"

is a special case of the distributive law for endless sums which was discussed at the end of the present chapter, but the proof forced us to think very carefully about endless sums.
    Potentially more serious is the apparently harmless step

$$1.111 \ldots - 0.111 \ldots = 1.$$

We have as yet not discussed the problem of performing arithmetic with infinite decimals. However we have stressed the importance of the *ordering* of the terms in endless sums, and if we rewrite 1.111 ... − 0.111 ... in full:

$$\left(1 + 1 \cdot \frac{1}{10} + 1 \cdot \frac{1}{10^2} + 1 \cdot \frac{1}{10^3} + \ldots\right) - \left(1 \cdot \frac{1}{10} + 1 \cdot \frac{1}{10^2} + 1 \cdot \frac{1}{10^3} + \ldots\right)$$

then we see that cancelling requires us to *shift* the term $-1 \cdot (1/10^i)$ in the second bracket so that it can cancel with the corresponding term in the first bracket. To convince you that this spells trouble in general try to make sense of the following three examples:

(a) Let $x = 1 + \dfrac{1}{2} + \dfrac{1}{3} + \dfrac{1}{4} + \dfrac{1}{5} + \ldots$

Multiply by 1/2

$$\frac{1}{2}x = \frac{1}{2} + \frac{1}{4} + \frac{1}{6} + \frac{1}{8} + \frac{1}{10} + \ldots$$

Subtract

$$\frac{1}{2}x = x - \frac{1}{2}x = \left(1 + \frac{1}{2} + \frac{1}{3} + \frac{1}{4} + \frac{1}{5} + \ldots\right) - \left(\frac{1}{2} + \frac{1}{4} + \frac{1}{6} + \frac{1}{8} + \ldots\right)$$

$$= 1 + \frac{1}{3} + \frac{1}{5} + \frac{1}{7} + \ldots$$

Thus

$$\frac{1}{2} + \frac{1}{4} + \frac{1}{6} + \frac{1}{8} + \ldots = \frac{1}{2} x = 1 + \frac{1}{3} + \frac{1}{5} + \frac{1}{7} + \ldots$$

(b)  Let $x = 1 - 1 + 1 - 1 + 1 - \ldots$
    Multiply by $-1$

$$-x = -1 + 1 - 1 + 1 - 1 + \ldots$$

Subtract

$$2x = x - (-x) = (1 + (-1 + 1 - 1 + 1 - \ldots))$$
$$- (-1 + 1 - 1 + 1 - 1 + \ldots) = 1$$

Hence

$$x = \frac{1}{2}$$

(c)  Let $x = 1 + 2 + 2^2 + 2^3 + \ldots$
Multiply by 2

$$2x = 2 + 2^2 + 2^3 + 2^4 + \ldots$$

Subtract

$$x = 2x - x = (2 + 2^2 + 2^3 + 2^4 + \ldots) - (1 + 2 + 2^2 + 2^3 + \ldots)$$

Hence

$$x = -1.$$

# Recurring Nines

We learn to compare ordinary finite decimals more or less *by eye :* thus .09 and .1 not only *look* different, they really *are* different—.09 being less than .1, since 9/100 (=.09) is less than 10/100 (=.1). But we have gone out of our way to stress the fact that, unlike ordinary finite decimals, *infinite decimals do not correspond to decimal fractions;* instead they have to be interpreted in a completely new way as *endless sums.* We therefore have to resist any temptation to assume that procedures which work with finite decimals will automatically carry over to infinite decimals.

Perhaps the most common indication that a student has failed to understand the crucial difference between finite and infinite decimals is the mistaken belief that

$$\text{just as} \quad .89 < .90$$
$$\text{and} \quad .99 < 1.0$$
$$\text{so also} \quad .8999 \ldots = .8\dot{9} < .9$$
$$\text{and} \quad .999 \ldots = .\dot{9} < 1.0.$$

The first hint that something unusual is going on in the case of *recurring nines* often arises when one works out the infinite decimals corresponding to 1/9, 2/9, 3/9, etc. and discovers that

$$\frac{1}{9} = .111 \ldots$$

$$\frac{2}{9} = .222 \ldots$$

$$\frac{3}{9} = .333 \ldots$$

$$\vdots$$

$$\frac{8}{9} = .888 \ldots$$

whereas

$$\frac{9}{9} = 1.0.$$

Since we are generally encouraged in mathematics to expect, and to look for, patterns, you might well have anticipated finding that

$$\frac{9}{9} = .999 \ldots$$

However, our *division process* for 9/9 clearly produces 1.0 and *not* .999 ... .

Nevertheless, our curiosity having been aroused, let us see what exactly we *can* say about the infinite decimal .999 ... . In the light of the previous chapter we realise that .999 ... must be interpreted as an *endless sum*

$$.999 \ldots = 9 \cdot \frac{1}{10} + 9 \cdot \frac{1}{10^2} + 9 \cdot \frac{1}{10^3} + 9 \cdot \frac{1}{10^4} + \cdots .$$

We also realise that the numerical value which we should assign to this endless sum is obtained from the corresponding *endless sequence of finite sums* :

$$9 \cdot \frac{1}{10}, \quad 9 \cdot \frac{1}{10} + 9 \cdot \frac{1}{10^2}, \quad 9 \cdot \frac{1}{10} + 9 \cdot \frac{1}{10^2} + 9 \cdot \frac{1}{10^3},$$

$$9 \cdot \frac{1}{10} + 9 \cdot \frac{1}{10^2} + 9 \cdot \frac{1}{10^3} + \frac{1}{10^4}, \ldots .$$

In the previous section we discovered that the behaviour of this endless sequence becomes much clearer if each finite sum of the form

$$a + ax + ax^2 + ax^3 + \ldots + ax^n$$

is rewritten in the form

$$\frac{a}{1-x} + \frac{a \cdot x^n}{x-1}$$

(provided $x$ is not equal to 1). In our case $a = 9 \cdot (1/10)$ and $x = 1/10$, so our endless sequence of finite sums can be rewritten as follows:

$$\frac{(9/10)}{1-(1/10)} + \frac{(9/10) \cdot (1/10)}{(1/10)-1}, \quad \frac{(9/10)}{1-(1/10)} + \frac{(9/10) \cdot (1/10^2)}{(1/10)-1},$$

$$\frac{(9/10)}{1-(1/10)} + \frac{(9/10) \cdot (1/10^3)}{(1/10)-1}, \quad \frac{(9/10)}{1-(1/10)} + \frac{(9/10) \cdot (1/10^4)}{(1/10)-1}, \ldots$$

which is really much simpler than it looks, since most of the awkward bits cancel to give

$$1 - \frac{1}{10}, \quad 1 - \frac{1}{10^2}, \quad 1 - \frac{1}{10^3}, \quad 1 - \frac{1}{10^4}, \ldots .$$

It is now clear that, although each of the finite sums

$$9 \cdot \frac{1}{10} + 9 \cdot \frac{1}{10^2} + 9 \cdot \frac{1}{10^3} + \ldots + 9 \cdot \frac{1}{10^n} =$$

$$\frac{(9/10)}{1-(1/10)} + \frac{(9/10) \cdot (1/10^n)}{(1/10)-1} = 1 - \frac{1}{10^n}$$

is strictly less than 1, the amount ($=1/10^n$) by which each finite sum differs from 1 shrinks rapidly towards zero as $n$ increases. *We thus have no choice but to assign the numerical value 1 to the endless sequence of finite sums, and hence also to the endless sum*

$$9 \cdot \frac{1}{10} + 9 \cdot \frac{1}{10^2} + 9 \cdot \frac{1}{10^3} + 9 \cdot \frac{1}{10^4} + \cdots .$$

However disturbing you may find this,[1] it is one of the unavoidable consequences of our decision

(i) firstly, to interpret infinite decimals as endless sums,

(ii) secondly, to interpret each endless sum in terms of the corresponding endless sequence of finite sums, and

(iii) thirdly, to choose the number which seems to be the *"endpoint"* towards which this endless sequence of finite sums is heading as the numerical value we assign *to the endless sum.*

This argument not only forces us to accept that

$$1.0 = .\dot{9} = .999\ldots$$

but applies also to *any infinite decimal which ends with recurring nines.* For example,

$$.41\dot{9} = .41999\ldots$$

has first to be interpreted as an endless sum

$$4 \cdot \frac{1}{10} + 1 \cdot \frac{1}{10^2} + 9 \cdot \frac{1}{10^3} + 9 \cdot \frac{1}{10^4} + 9 \cdot \frac{1}{10^5} + \cdots ,$$

which must in turn be interpreted in terms of the endless sequence of finite sums

$$4 \cdot \frac{1}{10}, \quad 4 \cdot \frac{1}{10} + 1 \cdot \frac{1}{10^2}, \quad 4 \cdot \frac{1}{10} + 1 \cdot \frac{1}{10^2} + 9 \cdot \frac{1}{10^3},$$

$$4 \cdot \frac{1}{10} + 1 \cdot \frac{1}{10^2} + 9 \cdot \frac{1}{10^3} + 9 \cdot \frac{1}{10^4}, \ldots .$$

We may bracket each finite sum

$$4 \cdot \frac{1}{10} + 1 \cdot \frac{1}{10^2} + 9 \cdot \frac{1}{10^3} + 9 \cdot \frac{1}{10^4} + \cdots + 9 \cdot \frac{1}{10^n}$$

---

[1] Many students insist—quite independently, and hopefully without having been taught—that

*1.0 and .$\dot{9}$ differ by .00 ... 1 = .$\dot{0}$1.*

Though one is bound to admire the imagination which invents .$\dot{0}$1, it is painfully clear what it is intended to mean—namely, *an infinite string of 0's, with a 1 in the last place,* which then adds to *the last 9 in .999 ... .* Unfortunately this misses the whole point of infinite decimals—namely, that, like the endless sequence of counting numbers, the sequence of digits of an infinite decimal *has no end,* and hence *no last place.*

into two parts—taking first all those terms before the recurring nines begin $(4 \cdot (1/10) + 1 \cdot (1/10^2) = (41/10^2))$, and then those terms involving the recurring nines $(9 \cdot (1/10^3) + 9 \cdot (1/10^4) + \ldots + 9 \cdot (1/10^n))$. The first part contributes a constant quantity (namely $41/10^2$) to every finite sum after the first, and the second part can be rewritten in the more enlightening form

$$\frac{9}{10^3} + \frac{9}{10^4} + \ldots + \frac{9}{10^n} = \frac{(9/10^3)}{1 - (1/10)} + \frac{(9/10^3) \cdot (1/10)^{n-2}}{(1/10) - 1} = \frac{1}{10^2} - \frac{1}{10^n}.$$

Thus our sequence of finite sums takes on the very simple form

$$4 \cdot \frac{1}{10}, \quad \frac{41}{10^2}, \quad \frac{41}{10^2} + \left(\frac{1}{10^2} - \frac{1}{10^3}\right),$$

$$\frac{41}{10^2} + \left(\frac{1}{10^2} - \frac{1}{10^4}\right), \quad \frac{41}{10^2} + \left(\frac{1}{10^2} - \frac{1}{10^5}\right), \ldots .$$

Though each of these terms in strictly less than $42/10^2$, they differ from $42/10^2$, by an amount $(=1/10^n)$ which shrinks rapidly toward zero. We therefore assign the numerical value $42/10^2$ to the sequence of finite sums, and hence to the original endless sum, whence we have

$$.41\dot{9} = \frac{42}{10^2} = .42.$$

EXERCISES

1. Find alternative decimal representations of the following:

   (i) 3.$\dot{9}$,     (ii) 99.$\dot{9}$,     (iii) 3419.$\dot{9}$,

   (iv) 3.2,     (v) 98,     (vi) 3400.

2. Show that the only positive real numbers which can be written as decimals in two distinct ways are the positive *decimal fractions* (that is, numbers which can be written in the form

$$\frac{a}{10^n}$$

where $a > 0$ and $n > 0$ are whole numbers). [Hint: Suppose a number can be written as a decimal in two distinct ways; then consider the first place in which the two expressions differ.]

# Fractions and Recurring Decimals

Each time we have worked out the infinite decimal corresponding to a fraction, the string of decimal digits has always ended with a *repeating block*: for example,

$$1/3 = .\dot{3}$$

$$8/70 = .1\dot{1}4285\dot{7}$$

$$1/13 = .\dot{0}7692\dot{3}$$

$$1/11 = .\dot{0}\dot{9}.$$

This is also true for ordinary *finite* decimals, which can either be thought of as ending with *repeated zeros*, or be written in the alternative form which was discussed in the previous section (that is, involving *recurring nines*): for example

$$1/2 = .5 = .5\dot{0} = .4\dot{9}$$

$$1/8 = .125 = .125\dot{0} = .124\dot{9}$$

$$2/5 = .4 = .4\dot{0} = .3\dot{9}.$$

In this chapter and its Exercises we shall tackle three questions which arise naturally from this observation.

(i) Firstly, is it true that a decimal which represents a fraction *has to be a recurring decimal*?

(ii) Secondly, does the relationship between fractions and recurring decimals also hold the other way round? That is, does every recurring decimal have to correspond to a fraction?

(iii) And finally, what is the exact relationship between
the fraction we start with,
the length of its repeating block, and
the digits which occur in that repeating block?

The answer to the first question should soon become clear to anyone who reads this paragraph and the next, and then works through Exercise 1. To help us analyse the process of calculating the decimal representation of a fraction *by division*, we shall consider the particular fraction 8/70.

$$
\begin{array}{r}
.1 \quad 1 \quad 4 \quad 2 \quad 8 \quad 5 \quad 7 \quad , \quad , \quad , \quad \ldots \\
70 \overline{\smash{\big)}\ 8.0^{1}0^{3}0^{2}0^{6}0^{4}0^{5}0^{1}0 \ 0 \ 0 \ \ldots}
\end{array}
$$

At the first step (*70 into 80*) we are still working directly with the two numbers (8 and 70) with which we started. But the second step (*70 into 100*), and each succeeding step, *is determined solely by the remainder from the previous step*. If at any stage we happen to obtain the remainder 0, then the division process stops, and the fraction we started with has been successfully expressed as an ordinary finite decimal, which we may think of as ending in *recurring zeros*: thus in such a case, the decimal we obtain to represent the fraction we started with is indeed a *recurring decimal*. However, in the case of 8/70 we can never obtain the remainder zero.[1] Thus, since we are dividing by 70, the only remainders we could possibly get are

$$1, 2, 3, \ldots, 69.$$

But if we look at the *seventy steps* between the *second* step (*70 into 100*) *where the string of zeros begins*, and the *seventy-first* step, we see that these use up *seventy* remainders; since there are at most sixty-nine different remainders available, we can be sure that some remainder gets repeated between the second and the seventy-first step.[2] But if the $i^{th}$ remainder is repeated later at the $j^{th}$ step, then the $i + 1^{th}$ remainder is automatically repeated at the $j + 1^{th}$ step, and the whole recurring process "takes off." In the case of 8/70, once the remainder 10 appears again at the eighth step, the eighth step (*70 into 100*) produces exactly the same result as the second step (*70 into 100*) and exactly the same remainder (namely *30*): thus we generate the recurring sequence of digits

$$1, 4, 2, 8, 5, 7, 1, 4, 2, 8, 5, 7, 1, 4, \ldots .$$

In exactly the same way, whatever fraction $m/n$ we start with, if the remainder zero occurs, then the division process transforms the fraction $m/n$ into an ordinary finite decimal (see Exercise 2); otherwise the division process transforms $m/n$ into an infinite decimal. In the latter case, since we are dividing by $n$, the only possible remainders which can occur are

$$1, 2, 3, \ldots, n - 1$$

so that in carrying out the division

$$n \overline{\smash{)}m . 0\ 0\ 0 \ldots}$$

we are bound to get a repeated remainder between the step involving the first recurring zero of the dividend, and the step involving the $n^{th}$ recurring zero; this repeated remainder then sets the whole recurring process in motion, and we obtain $m/n$ as a *recurring decimal*. Moreover, since in this

---

[1] We can always tell at the start whether the particular fraction we are working with can be expressed as a *decimal fraction* $a/10^n$, and thus as an ordinary (finite) decimal (see Exercise 2).

[2] In this particular case things are actually much simpler than this, though the principle is exactly the same; for once the zeros begin, we are actually dividing 70 into multiples of 10, so the only possible remainders are 10, 20, 30, 40, 50, 60. Thus we can in fact be sure that some remainder gets repeated between the second and the eighth step: in fact the remainder 10 at the eighth step is equal to the remainder at the second step.

case $m/n$ is not a decimal fraction, this is the only possible way of writing $m/n$ as a decimal (see Chapter II.8, Exercise 2). Thus not only does the division process transform each (positive) fraction $m/n$ into a *recurring decimal*, but *every decimal which represents a fraction*[3] *has to be a recurring decimal*. To reinforce this conclusion and to prepare for what follows you should complete Exercise 1 before going any further.

*        *        *        *        *

And so we come to the second of our three questions about infinite recurring decimals and fractions:

(ii) *Does every recurring decimal necessarily correspond to a fraction?*

We shall be as brief as we can. Suppose I write down the first recurring decimal that comes into my head:

$$123.4567897897897897 \ldots .$$

In looking for a number which this represents, we first interpret it as an endless sum

$$123.456 + \frac{789}{10^6} + \frac{789}{10^9} + \frac{789}{10^{12}} + \frac{789}{10^{15}} + \ldots .$$

If we then consider the corresponding endless sequence of finite sums, we discover as in Chapter II.7 that

$$123.456 + \left( \frac{789}{10^6} + \frac{789}{10^9} + \ldots + \frac{789}{10^{3(n+1)}} \right) =$$

$$\frac{123456}{10^3} + \left( \frac{789}{10^3(10^3 - 1)} - \frac{789}{10^{3n}(10^3 - 1)} \right)$$

so we assign to the endless sum the numerical value

$$\frac{123456}{10^3} + \frac{789}{10^3(10^3 - 1)}$$

*which is clearly a fraction*. The same argument works in every case: in general a recurring decimal has the form

$$c_1 \ldots c_n . b_1 \ldots b_m \dot{a}_1 \ldots \dot{a}_k ,$$

where the $a$'s, $b$'s and $c$'s are single decimal digits from 0–9. This corresponds to the endless sum

$$c_1 \ldots c_n . b_1 \ldots b_m + \frac{a_1 \ldots a_k}{10^{m+k}} \left( 1 + \frac{1}{10^k} + \frac{1}{10^{2k}} + \ldots \right)$$

---

[3] Whether it is obtained by the division process (like $2/5 = .4$) or not (like $2/5 = .3\dot{9}$).

to which we assign the value

$$\frac{c_1 \ldots c_n b_1 \ldots b_m}{10^m} + \frac{a_1 \ldots a_k}{10^{m+k}} \cdot \frac{10^k}{10^k - 1}.$$

*Thus any recurring decimal with a repeating block of length k corresponds to a fraction whose denominator has the form*

$$10^m \cdot 10^k - 1.$$

(It follows from the first of our three questions that *every* fraction $a/b$ can be written in the form

$$\frac{c}{10^m(10^k - 1)}$$

—see Exercise 7.)

In contrast to our complete solution of the first two of our three questions about the relationship between infinite recurring decimals and fractions, we shall scarcely begin to answer even the first part of our final question:

(iii) *What is the exact relationship between*
  *the fraction m/n we start with,*
  *the length of the repeating block in its decimal, and*
  *the digits which occur in the repeating block?*

Indeed almost all our efforts will go into trying to understand

*What determines the length of the repeating block?*

As a first step, we observe that it is the *value* of the fraction $m/n$ rather than the particular values of $m$ and $n$ which determine the decimal (and hence the length of the repeating block). Thus it is natural to cancel any common factors of $m$ and $n$; in other words, *we may assume that*

$$hcf(m, n) = 1.$$

Observe next that the recurring block in

$$9/7 = 1.\dot{2}8571\dot{4}$$

is exactly the same as the recurring block in

$$2/7 = .\dot{2}8571\dot{4}$$

since

$$\frac{9}{7} = 1 + \frac{2}{7}.$$

In exactly the same way we can always remove the integer part of $m/n$ and concentrate on the strictly fractional part: in other words, *we can assume that $m < n$.*

Fractions giving rise to finite decimals can be considered as a special case of fractions $m/n$ whose decimal has a *recurring block of length one*. If the decimal for $m/n$ ends in recurring zeros or recurring nines, then $m/n$ must be an ordinary decimal fraction—that is there exist numbers $a$, $b$ for which

$$\frac{m}{n} = \frac{a}{10^b} \, ;$$

this will happen whenever $n$ is a product of 2's and 5's only:

$$n = 2^p \cdot 5^q \text{ (see Exercise 2).}$$

If the decimal for $m/n$ ends in recurring ones, twos, threes, fours, fives, sixes, sevens, or eights, then $m/n$ must be of the form

$$\frac{m}{n} = \frac{a}{9.10^b}$$

where $a$ is not a multiple of 9 (if $a$ is a multiple of 9 then we are back in the previous case); hence $n$ must be of the form

$$n = 3 \cdot 2^p \cdot 5^q \quad \text{or} \quad 3^2 \cdot 2^p \cdot 5^q.$$

Combining the two cases we see that

*$m/n$ has a recurring block of length one precisely when $n$ divides $9.10^b$* $(b \geq 0)$;

for further details see Exercise 3. Thus, if we wish, we may assume that $m/n$ has an infinite decimal with a repeating block of length at least two.

In the hope of simplifying the problem further we next try to decide whether it is the numerator $m$, the denominator $n$, or some combination of the two which determines the length of the recurring block. From Exercise 1 you will have noticed that

$$1/7 = .\dot{1}4285\dot{7}$$

$$2/7 = .\dot{2}8571\dot{4}$$

$$3/7 = .\dot{4}2857\dot{1}$$

$$.$$

$$.$$

$$6/7 = .\dot{8}5714\dot{2}$$

$$(7/7 = 1.0)$$

which suggests that the numerator is only relevant insofar as it may cancel out part of the denominator[4]—in which case $n$ is no longer the *true* denominator. But is it true in general that, *if the fraction $m/n$ is in its lowest terms, then the numerator has no effect on the length of the repeating block?*

---

[4] But beware of the temptation to conclude that the digits involved in the repeating block only depend on the denominator: for you should also have noticed from Exercise 1 that
$$1/11 = .\dot{0}\dot{9}, \quad 2/11 = .\dot{1}\dot{8}, \dots ; \quad \text{and} \quad 1/13 = .\dot{0}7692\dot{3}, \quad 2/13 = .\dot{1}5384\dot{6}, \dots .$$

We shall now show, by comparing $m/n$ with $1/n$, that this really is the case. But consider first the specific example $n = 7$, $m = 3$: this has been written out in a way which should clarify the proof for the general case—in particular we have only allowed ourselves to perform those calculations which can also be carried out in the general case. First of all $1/7$ has a repeating block of length six:

$$\frac{1}{7} = .14285714285714 \ldots .$$

From this it follows that

$$.142857 < \frac{1}{7}$$

that is

$$\frac{142857}{10^6} < \frac{1}{7}$$

so

$$3 \times \frac{142857}{10^6} < 3 \times \frac{1}{7} < 1.$$

Hence, in particular,

$$3 \times 142857 < 10^6;$$

so

$$3 \times 142857 = b_1 b_2 b_3 b_4 b_5 b_6$$

for some choice of digits $b_i$ from 0-9. Hence

$$\frac{3}{7} = 3 \times \frac{1}{7} = 3 \times .14285714285714 \ldots$$
$$= \quad .b_1 b_2 b_3 b_4 b_5 b_6 b_1 b_2 b_3 b_4 b_5 b_6 b_1 b_2 \ldots {}^5$$
$$= \quad .\dot{b}_1 b_2 b_3 b_4 b_5 \dot{b}_6 .$$

Thus $3/7$ also has a repeating block of length six.

The proof for $m/n$ in general (assuming only that $m < n$) is almost exactly the same. Suppose first of all that $1/n$ has a repeating block of length $k$:

$$\frac{1}{n} = .a_1 a_2 a_3 \ldots a_k a_1 a_2 a_3 \ldots a_k a_1 a_2 \ldots$$

---

[5] We have here made use of the *distributive law for endless sums*. That this is permissible follows from the discussion at the end of Chapter II.7.

where the $a_i$'s are all single digits from 0–9. Then it follows that

$$.a_1 a_2 a_3 \ldots a_k < \frac{1}{n};$$

that is

$$\frac{a_1 a_2 a_3 \ldots a_k}{10^k} < \frac{1}{n}.$$

Thus

$$m \times \frac{a_1 a_2 a_3 \ldots a_k}{10^k} < m \times \frac{1}{n} < 1;$$

so

$$m \times a_1 a_2 a_3 \ldots a_k < 10^k$$

and

$$m \times a_1 a_2 a_3 \ldots a_k = b_1 b_2 b_3 \ldots b_k$$

for some choice of digits $b_i$ from 0–9. Hence

$$\frac{m}{n} = m \times \frac{1}{n} = m \times .a_1 a_2 a_3 \ldots a_k a_1 a_2 a_3 \ldots a_k a_1 a_2 \ldots$$

$$= .b_1 b_2 b_3 \ldots b_k b_1 b_2 b_3 \ldots b_k b_1 b_2 \ldots .$$

Thus $m/n$ also has a repeating block of length $k$.

You may have noticed that we have up to now made no use of the important extra condition

$$hcf(m, n) = 1.$$

As a result, though we have proved that $m/n$ has *a* repeating block of length $k$, there is no reason to believe that it is necessarily *the shortest* repeating block. For example if $n = 21$, $m = 7$, then

$$\frac{1}{21} = .0\dot{4}761\dot{9}$$

has a repeating block of length six, and

$$\frac{7}{21} = .\dot{3}3333\dot{3}$$

also has a repeating block of length six; but it is not the shortest repeating block. In general, if

$$\frac{m}{n} = .\dot{b}_1 b_2 b_3 \ldots \dot{b}_k$$

and $c_1 c_2 \ldots c_j$ is the shortest repeating block for $m/n$, then $b_1 b_2 \ldots b_k$ is simply made up of a number of copies of $c_1 c_2 \ldots c_j$. So far we have only proved the following result:

*If the shortest repeating block of $1/n$ has length $k$*
*then the length of the shortest repeating block of $m/n$ is a factor of $k$.*

In Exercise 3 you have the chance to prove that

*If the shortest repeating block of $1/n$ has length $k$, and*
*if $hcf(m, n) = 1$,*
*then the shortest repeating block of $m/n$ has length exactly $k$.*

All the information we require is therefore contained in the decimal representations of fractions of the form $1/n$; so once we have answered our question (iii) for the recurring decimal representations of fractions of this form we shall be more or less done.

But, as yet, we have not the slightest idea how the denominator $n$ is related either to the length of the repeating block, or to the digits which occur within the repeating block. Hence, before we start to think about proving anything at all, we shall investigate particular values of $n$ in search of some clue as to *what exactly we should be trying to prove.* And rather than calculating the infinite decimal representation of $1/n$ for values of $n$ chosen at random, we shall set about collecting *systematic experimental data*—in other words, we shall make a complete list of the decimal representation of $1/n$ for values of $n$ up to some suitable point.[6]

For $n = 1$ we scarcely need to divide to get $\quad 1/1 = 1.$
For $n = 2$ we must work from scratch to get $\quad 1/2 = .5.$
For $n = 3$ we must work from scratch to get $\quad 1/3 = .\dot{3}.$
For $n = 4$ we can divide $1/2 = .5$ by 2 to get $\quad 1/4 = .25.$
For $n = 5$ we must work from scratch to get $\quad 1/5 = .2.$
For $n = 6$ we can divide $1/3 = .\dot{3}$ by 2 to get $\quad 1/6 = .1\dot{6}.$
For $n = 7$ we must work from scratch to get $\quad 1/7 = .\dot{1}4285\dot{7}.$
For $n = 8$ we can divide $1/4 = .25$ by 2 to get $\quad 1/8 = .125.$
For $n = 9$ we can divide $1/3 = .\dot{3}$ by 3 to get $\quad 1/9 = .\dot{1}.$
For $n = 10$ we can divide $1/5 = .2$ by 2 to get $\quad 1/10 = .1.$
For $n = 11$ we must work from scratch to get $\quad 1/11 = .\dot{0}\dot{9}.$
For $n = 12$ we can divide $1/6 = .1\dot{6}$ by 2 to get $\quad 1/12 = \underline{\qquad}.$
For $n = 13$ we must work from scratch to get $\quad 1/13 = \underline{\qquad}.$
For $n = 14 \ldots$

---

[6] Since we are not merely interested in getting "the right answer" but are trying to *analyse what exactly is going on,* there is little point simply using our own memory, or reciprocal tables, or the reciprocal key on a calculator. We shall therefore work everything out from scratch—at least until it is clear how tables or a machine can be best used.

Before going any further you should extend this table as far as $n = 29$, making a special note of those values of $n$ for which the decimal has to be worked out from scratch.

*       *       *       *       *

We could extend this list as far as we like, but we should perhaps pause for a moment to reflect on what seems to be happening. Firstly, the only cases you had to work out from scratch should have been

$$\frac{1}{17}, \quad \frac{1}{19}, \quad \frac{1}{23}, \quad \frac{1}{29};$$

all other values of $n$ between 14 and 29 can be factorised, say

$$n = a \cdot b$$

whence the decimal for $1/n$ can be calculated by dividing the decimal for $1/a$ by $b$.[7] Thus the decimal representations of numbers of the form

$$\frac{1}{p}, \qquad \text{where } p \text{ is a prime,}$$

are in some sense more basic than the others, and we may reasonably hope that once we understand what determines the length of the repeating block in these cases, then we should be able to use this to understand the other cases as well (see Exercise 4). We shall therefore restrict our attention from now on to the case where $n$ is a prime; and as a permanent reminder that we are dealing with a prime, we shall write $p$ in place of $n$.

The data so far collected on the length $k$ of the shortest repeating blocks for fractions of the form $1/p$, $p$ a prime, can be tabulated as follows:

| $k = 1$ | $k = 2$ | $k = 6$ | $k = 16$ | $k = 18$ | $k = 22$ | $k = 28$ |
|---|---|---|---|---|---|---|
| $1/2 = .5$<br>$1/3 = .\dot{3}$<br>$1/5 = .2$ | $1/11 = .\dot{0}\dot{9}$ | $1/7$<br>$1/13$ | $1/17$ | $1/19$ | $1/23$ | $1/29$ |

Though the cases

$$k = 6 \ \text{ for } \ \frac{1}{7}, \qquad k = 16 \ \text{ for } \ \frac{1}{17}, \qquad k = 18 \ \text{ for } \ \frac{1}{19},$$

$$k = 22 \ \text{ for } \ \frac{1}{23}, \qquad k = 28 \ \text{ for } \ \frac{1}{29}$$

---

[7] Or by dividing the decimal for $1/b$ by $a$, though in general it is easiest if one divides by the smallest possible factor of $n$.

are very suggestive of some connection between $k$ and $p - 1$, we must also account for the cases

$$k = 6 \quad \text{for} \quad \frac{1}{13}, \qquad k = 2 \quad \text{for} \quad \frac{1}{11}, \qquad k = 1 \quad \text{for} \quad \frac{1}{5},$$

$$k = 1 \quad \text{for} \quad \frac{1}{3}.$$

Moreover, before we get carried too far in the wrong direction we should remark that the reasoning on which we based our decision to focus attention on fractions of the form

$$\frac{1}{p} \quad \text{where } p \text{ is a prime}$$

had nothing whatever to do with the fact that we were expecting to work in base 10. It follows that the conclusion we eventually reach about "the exact relationship between $p$ and the shortest repeating block in the decimal for $1/p$" should be equally applicable to any other base. In seeking to discover this "exact relationship" we should therefore vary not only the prime $p$ as we have done above, but also the base with respect to which $1/p$ is expressed. As particular examples we shall consider base 9 and base 11.

   Let us therefore express our fractions $1/p$ in base 9. The cases $p = 2, 3, 5$ are left as exercises, but we shall work through the case $p = 7$. To express $1/7$ in base 9 we may first write the dividend 1 in base 9 and then carry out the division

$$7 \overline{)1 \,.\, 0\,0\,0\,0\,\ldots}$$

The first step (*7 into* $10_{base\ 9}$) gives "*1 remainder 2*" so we obtain

$$\begin{array}{r} .\,1 \\ \hline 7 \overline{)1 \,.\, 0^2 0\,0\,0} \end{array}$$

The second step (*7 into* $20_{base\ 9}$) gives "*2 remainder 4*" so we obtain

$$\begin{array}{r} .\,1\,2 \\ \hline 7 \overline{)1 \,.\, 0^2 0^4 0\,0\,\ldots} \end{array}$$

The third step (*7 into* $40_{base\ 9}$) gives "*5 remainder 1*" so we obtain

$$\begin{array}{r} .\,1\,2\,5 \\ \hline 7 \overline{)1 \,.\, 0^2 0^4 0^1 0\,\ldots} \end{array}$$

and the recurring process has begun. Thus in base 9 the fraction for $1/7$ has shortest repeating block of length 3. You should now express all the following fractions in base 9:[8]

$$\frac{1}{2} = \underline{\hspace{3cm}} \text{ base 9} \qquad \frac{1}{13} = \underline{\hspace{3cm}} \text{ base 9}$$

$$\frac{1}{3} = \underline{\hspace{3cm}} \text{ base 9} \qquad \frac{1}{17} = \underline{\hspace{3cm}} \text{ base 9}$$

$$\frac{1}{5} = \underline{\hspace{3cm}} \text{ base 9} \qquad \frac{1}{19} = \underline{\hspace{3cm}} \text{ base 9}$$

$$\frac{1}{7} = \underline{\hspace{3cm}} \text{ base 9} \qquad \frac{1}{23} = \underline{\hspace{3cm}} \text{ base 9}$$

$$\frac{1}{11} = \underline{\hspace{3cm}} \text{ base 9} \qquad \frac{1}{29} = \underline{\hspace{3cm}} \text{ base 9}$$

$$*\quad *\quad *\quad *\quad *$$

If $k$ is now used to denote the length of the shortest repeating block in *base 9*, then this information can be tabulated as before:

| $k = 1$ | $k = 2$ | $k = 3$ | $k = 5$ | $k = 8$ | $k = 9$ | $k = 11$ | $k = 14$ |
|---------|---------|---------|---------|---------|---------|----------|----------|
| 1/2 1/3 | 1/5 | 1/7 1/13 | 1/11 | 1/17 | 1/19 | 1/23 | 1/29 |

The *base 10* list seemed to suggest that $k = p - 1$ except for small values of $p$; but we see from the *base 9* list that this view is untenable. However we still find that

$$k \text{ divides } p - 1$$

in every case. To test this guess we should not only work out $1/p$ for primes

---

[8] A purist would no doubt claim that the whole calculation should be carried out strictly in *base 9*: thus for 1/17 we would write $17 = 18_{base\ 9}$ and proceed as follows:

$$\begin{array}{r} .0\ \ 4\ \ 6\ \ 7\ \ \ 8\ 4\ 2\ 1\ 0 \dots \\ 18\,\overline{)\,1\,.\,0^{10}0^{14}0^{16}0^{17}0^{8}0^{4}0^{2}0^{1}0 \dots} \end{array}$$

However this confuses me horribly, and I have to admit that I work unashamedly in *base 10*: thus for 1/17 I write

$$\begin{array}{r} .0\ \ \ 4\ \ \ 6\ \ \ 7\ \ \ 8\ \ \ 4\ \ 2\ \ 1\ \ \ 0 \\ 17\,\overline{)\,1\,.\,0\ \ \ 0\ \ \ 0\ \ \ 0\ \ \ 0\ \ \ 0\ \ 0\ \ 0\ \ \ 0 \dots} \\ 9\ \ \ 81\ \ 117\ 135\ 144\ \ 72\ \ 36\ \ 18 \\ \underline{68\ \ 102\ 119\ 136\ \ 68\ \ 34\ \ 17} \\ 13\ \ \ 15\ \ \ 16\ \ \ 8\ \ \ 4\ \ \ 2\ \ \ 1 \end{array}$$

$p > 29$, in *base 10* and *base 9*, but also repeat our calculations in another base—say *base 11*.[9] The reader should fill in the gaps in the following list:

$$\frac{1}{2} = .\dot{5}_{base\ 11} \qquad\qquad \frac{1}{13} = .\underline{\quad\quad\quad\quad}_{base\ 11}$$

$$\frac{1}{3} = .\dot{3}\dot{7}_{base\ 11} \qquad\qquad \frac{1}{17} = .\underline{\quad\quad\quad\quad\quad\quad}_{base\ 11}$$

$$\frac{1}{5} = .\dot{1}\dot{7}_{base\ 11} \qquad\qquad \frac{1}{19} = .\underline{\quad\quad\quad\quad\quad}_{base\ 11}$$

$$\frac{1}{7} = .\dot{1}6\dot{3}_{base\ 11} \qquad\qquad \frac{1}{23} = .\underline{\quad\quad\quad\quad}_{base\ 11}$$

$$\frac{1}{11} = .\dot{1}_{base\ 11} \qquad\qquad \frac{1}{29} = .\underline{\quad\quad\quad\quad}_{base\ 11}$$

*     *     *     *     *

If we use $k$ again to denote the length of the shortest repeating block but this time in *base 11*, then we obtain the following table:

| $k = 1$ | $k = 2$ | $k = 3$ | $k = 16$ | $k = 22$ | $k = 28$ |
|---------|---------|---------|----------|----------|----------|
| 1/2 1/11 | 1/3 1/5 | 1/7 1/19 | 1/17 | 1/23 | 1/29 |

So it looks as if what we guessed is in fact true—though it is presumably not the whole truth. But we have stumbled on no apparent reason why it should be true, nor any indication how we should go about proving it (see Exercises 5 and 6). Nevertheless, if $p$ is a prime which does not divide $m$, then we feel fairly confident that

*if $m/p$ is expressed in any base, then the length of the shortest repeating block always divides $p - 1$.*

However any further investigation either into the precise length of the repeating block and the digits it contains, or into the general case $1/n$ where $n$ is not a prime, must be left in the hands of the exercises and the reader's own curiosity.

---

[9] To write numbers in *base 11* we need not only the digits 0–9 but also an extra symbol, say $t$, to represent *ten*. Thus

$$\text{"ten"} = t_{base\ 11}, \text{ "one hundred and ten"} = t0_{base\ 11}, \frac{10}{11} = .t_{base\ 11}$$

EXERCISES

1. If we use the division process to divide 1 by 7

$$0 . 1 \ 4 \ 2 \ 8 \ 5 \ 7 \ 1 \ 4 ..$$
$$7 \overline{| 1 . \ {}^1 0 \ {}^3 0 \ {}^2 0 \ {}^6 0 \ {}^4 0 \ {}^5 0 \ {}^1 0 \ {}^3 0 ..}$$

then we generate not only the decimal representation for $1/7$:

$$1/7 = 0.14285714 \ldots$$

but also the sequence of *remainders to be carried*

$$1, \quad 3, \quad 2, \quad 6, \quad 4, \quad 5, \quad 1, \quad 3, \quad \ldots \quad .$$

Use the division process to find (a) the decimal representation for each of the following fractions, and (b) the sequence of remainders:

(i) $1/7 = .14285714 \ldots$ ; remainders      $1, 3, 2, 6, 4, 5, 1, 3, \ldots$
(ii) $2/7 =$
(iii) $3/7 =$
(iv) $4/7 =$
(v) $5/7 =$
(vi) $6/7$
(vii) $9/7$
(viii) $1/11 =$
(ix) $2/11 =$
(x) $3/11 =$
(xi) $1/13 =$
(xii) $2/13 =$
(xiii) $3/13 =$

2. (i) Use the division process to express $1/32$ as a decimal.

(ii) Show that if $m, n$ are positive whole numbers with

$$hcf(m, n) = 1,$$

and

$$\frac{m}{n} = .a_1 a_2 \ldots a_k$$

is an ordinary (finite) decimal, where the $a_i$ are ordinary decimal digits from 0–9, then $n$ divides $10^k$ exactly.

(iii) Show that if $n$ divides $10^k$ (that is $10^k = n \times c$ for some $c$), then

$$\frac{m}{n} = .a_1 a_2 \ldots a_k$$

where $a_1 a_2 \ldots a_k$ is the *base 10* representation of $m \times c$.

(iv) Use (iii) to write the decimal for $1/16$ without using the division process.

3. Suppose we wish to express $.2\dot{1}$ as a fraction. Then

$$.2\dot{1} = .2 + .0\dot{1}$$

$$= \frac{2}{10} + \left( \frac{1}{10} \times .\dot{1} \right)$$

$$= \frac{2}{10} + \left( \frac{1}{10} \times \frac{1}{9} \right) = \frac{19}{90} .$$

Use similar methods to express each of the following as a fraction:

(i) $.6\dot{1}$                                    (iv) $.67\dot{8}$

(ii) $.7\dot{2}$                                    (v) $.5678\dot{9}$

(iii) $.34\dot{3}$                                   (vi) $.12345\dot{6}$

4. In Chapter II.3 we used the game of Euclid to introduce a procedure for finding the highest common factor of two whole numbers $m$, $n$. For $m = 5776$, $n = 9633$ we transformed the pair $(5776, 9633)$ thus:

$(5776, 9633)$
$\downarrow$          $9633 = 1 \cdot 5776 + 3857$ (where $0 < 3857 < 5776$)
$(5776, 3857)$
$\downarrow$          $5776 = 1 \cdot 3857 + 1919$ (where $0 < 1919 < 3857$)
$(1919, 3857)$
$\downarrow$          $3857 = 2 \cdot 1919 + 19$   (where $0 < 19 < 1919$)
$(1919, \quad 19)$
$\downarrow$          $1919 = 101 \cdot 19 + 0$
$(0, \qquad 19)$

At each stage we produce a new pair whose common factors are precisely the same as the common factors of the previous pair (and hence also of the original pair): thus

$$19 = hcf(5776, 9633).$$

(i) Find whole numbers $a$, $b$ such that

$$19 = a \cdot 5776 + b \cdot 9633.$$

[Hint: Start with the last but one equation $3857 = 2 \cdot 1919 + 19$; then substitute for 1919 in terms of 5776 and 3857 from the previous equation; ... .]

(ii) In general, given two positive whole numbers $m$, $n$ we can find $hcf(m, n)$ by carrying out a similar sequence of steps until we eventually reach a pair of the form $(0, r_k)$, in which case $r_k = hcf(m, n)$: thus

$(m, n)$
$\downarrow$                          $m = q_1 \cdot n + r_1$ (where $0 < r_1 < n$)
$(m - q_1 \cdot n, n) = (r_1, n)$
$\downarrow$                          $n = q_2 \cdot r_1 + r_2$ (where $0 < r_2 < r_1$)
$(r_1, n - q_2 \cdot r_1) = (r_1, r_2)$
$\downarrow$                          $r_1 = q_3 \cdot r_2 + r_3$ (where $0 < r_3 < r_2$)
$\vdots$
$\downarrow$                          $r_{k-2} = q_k \cdot r_{k-1} + r_k$ (where $0 < r_k < r_{k-1}$)
$(r_{k-1}, r_{k-2} - q_k \cdot r_{k-1}) = (r_{k-1}, r_k)$
$\downarrow$                          $r_{k-1} = q_{k+1} \cdot r_k + 0.$
$(0, r_k)$

Show that there exist whole numbers $a$, $b$ such that

$$r_k = a \cdot m + b \cdot n.$$

[Hint: Start with the equation

$$r_k = r_{k-2} - q_k \cdot r_{k-1};$$

then substitute for $r_{k-1}$ in terms of $r_{k-2}$ and $r_{k-3}$ from the previous equation—working backwards until one eventually substitutes for $r_2$ in terms of $r_1$ and $n$ from the equation $n = q_2 \cdot r_1 + r_2$, and finally substitutes for $r_1$ in terms of $m$ and $n$ from the equation $m = q_1 \cdot n + r_1$.]

(iii) Let $m$, $n$ be positive whole numbers with

$$hcf(m, n) = 1.$$

Then by part (ii) there exist whole numbers $a$, $b$ with

$$1 = a \cdot m + b \cdot n.$$

Thus

$$\frac{1}{n} = a \cdot \frac{m}{n} + b.$$

Deduce that the *lengths* of the shortest repeating blocks in the decimals for $1/n$ and for $m/n$ are the same.

*In the remaining exercises we tackle the question of the length of the shortest recurring block of decimal digits in the decimal for $1/p$ when $p$ is a prime number.*

5. (i) Look at the table below. It is supposed to go on for ever.

| 0 | 1 | 2 | 3 | 4 | 5 | 6 |
|---|---|---|---|---|---|---|
| 0 | 1 | 2 | 3 | 4 | 5 | 6 |
| 7 | 8 | 9 | 10 | 11 | 12 | 13 |
| 14 | 15 | 16 | 17 | 18 | 19 | 20 |
| 21 | 22 | 23 | 24 | 25 | 26 | 27 |
| 28 | 29 | 30 | 31 | 32 | 33 | 34 |
| 35 | 36 | 37 | 38 | 39 | 40 | 41 |
| 42 | 43 | . | . | . | . | . |
| . | . | . | | | | |

(a) How could you describe all the numbers in column 0 *in a single phrase*?

(b) Give an equally short description of all the numbers in column 1.

(c) Do the same for column 2; and for column 3; ... ; and for column 6.

(d) Which columns contain squares?

(e) Which columns contain cubes?

(f) I am thinking of a multiple of 7. In which column must its square lie? Explain.

(g) Suppose I put $r_0 = 1$: in which column will I find $10 \cdot r_0$? (And in which column will I find 10?)

   Suppose I then put $r_1 = 3$: in which column will I find $10 \cdot r_1$? (And in which column will I find $10^2$?)

   Suppose I then put $r_2 = 2$: in which column will I find $10 \cdot r_2$? (And in which column will I find $10^3$?)

   Suppose I then put $r_3 = 6$: in which column will I find $10 \cdot r_3$? (And in which column will I find $10^4$?)

   Suppose I then put $r_4 = 4$: in which column will I find $10 \cdot r_4$? (And in which column will I find $10^5$?)

Suppose I then put $r_5 = 5$: in which column will I find $10 \cdot r_5$? (And in which column will I find $10^6$?)

Suppose I then put $r_6 = 1$: ...

Etc.

(h) You may not have recognised the procedure given in part (g) as describing exactly the division process for finding the decimal for 1/7. It is camouflaged by the fact that it focusses attention on *the remainders* rather than *the decimal digits* themselves.

(ii) Before considering the general case you should work through one more example.

Imagine all the whole numbers placed in thirteen columns.

| 0 | 1 | 2 | 3 | 4 | 5 | 6 | 7 | 8 | 9 | 10 | 11 | 12 |
|---|---|---|---|---|---|---|---|---|---|----|----|----|
| 13 | 14 | 15 | 16 | 17 | 18 | 19 | 20 | 21 | 22 | 23 | 24 | 25 |
| 26 | 27 | 28 | 29 | 30 | 31 | 32 | 33 | 34 | 35 | 36 | 37 | 38 |
| 39 | 40 | . | . | . | . | . | . | . | . | . | . | . |
| . | . | . | | | | | | | | | | |

Use these to help analyse the remainders which arise when we divide 1 by 13

$$\begin{array}{r} 0 . . . . . . \\ \hline 13 \,\big|\, 1 . 0\,0\,0\,0\,0 \dots \end{array}$$

The first step (*13 into 1*) gives remainder $r_0 = 1$: $10 \cdot r_0$ lies in column _____ $= r_1$. (In which column is $10^1$?)

$r_1$ is now the remainder from the second step (*13 into 10*): $10 \cdot r_1$ lies in column _____ $= r_2$. (In which column is $10^2$?)

$r_2$ is now the remainder from the third step (*13 into 30*): $10 \cdot r_2$ lies in column _____ $= r_3$. (In which column is $10^3$?)

$r_3$ is now the remainder from the fourth step (*13 into 20*): $10 \cdot r_3$ lies in column _____ $= r_4$. (In which column is $10^4$?)

$r_4$ is now the remainder from the fifth step (*13 into 60*): $10 \cdot r_4$ lies in column _____ $= r_5$. (In which column is $10^5$?)

$r_5$ is now the remainder from the sixth step (*13 into 40*): $10 \cdot r_5$ lies in column _____ $= r_6$. (In which column is $10^6$?)

(iii) Let $p$ be any prime number, and consider the process of dividing 1 by $p$ in *base 10*.

In parts (i)(g) and (ii) it was hinted rather strongly that if the whole numbers are put into $p$ columns (column 0 to column $p-1$), then the length of the shortest repeating block of remainders is equal to

*the smallest power $k$ for which $10^k$ is in column 1.*

We shall now prove, by induction on $n$, that in the process of dividing 1 by $p$

*if $r_{n-1}$ is the remainder at the $n^{th}$ step of the division,*
*then $10^n$ and $r_n$ occur in the same column.*

(a) At the first step ($p$ into $1$) we have remainder $r_0 = 1$. Clearly $10^0$ and $r_0$ occur in the same column (column 1).

At the second step ($p$ into $10 \cdot r_0$) we have remainder $r_1$; that is,

$$10 = q_1 \cdot p + r_1 \qquad \text{where } 0 \le r_1 < p.$$

This simply says that $10^1$ and $r_1$ occur in the same column. At the third step ($p$ into $10 \cdot r_1$) we have remainder $r_2$. Show that $10^2$ and $10 \cdot r_1$ occur in the same column. Deduce that $10^2$ and $r_2$ occur in the same column.

(b) (Induction step). Suppose that for some value of $n$ we have proved that $10^{n-1}$ and $r_{n-1}$ occur in the same column.

Deduce that $10 \cdot 10^{n-1}$ and $10 \cdot r_{n-1}$ occur in the same column.

At the $n + 1^{\text{st}}$ step of the division process ($p$ into $10 \cdot r_{n-1}$) we obtain the remainder $r_n$; that is

$$10 \cdot r_{n-1} = q_n \cdot p + r_n \qquad \text{where } 0 \le r_n < p.$$

This simply says that $10 \cdot r_{n-1}$ and $r_n$ occur in the same column.

Deduce that $10^n$ and $r_n$ occur in the same column.

### Recurring Decimals—The Story So Far

The time has come for us to review our progress. We started by investigating the following question:

(1) What determines the length of the shortest recurring block of decimal digits in a decimal representation of the arbitrary fraction $m/n$?

We showed first that we could assume

(2) $hcf(m, n) = 1$.

We also showed that the general solution depended only on $n$, so that we could safely assume

(3) $m = 1$.

We then argued that in order to resolve the general case, we should first try to understand the special case in which

(4) $n = p$ is a prime number.

At this stage we remarked that whatever result we expected to prove about the length of the shortest recurring block for $1/p$ should be equally true in any base. The evidence gathered suggested that we should try to prove:

*The length $k$ of the shortest recurring block for $1/p$ (in any base) always divides $p - 1$.*

We then reinterpreted this number $k$ as being precisely equal to

*the length of the shortest repeating block of remainders which occur in the process of dividing 1 by p.*

For *base 10* we showed in Exercise 5 that if the whole numbers are written out in $p$ columns (column 0 to column $p - 1$), then the required number is equal to

*the smallest power $k$ for which $10^k$ occurs in column 1.*

A similar argument applies in any base $b$ which is not a multiple of $p$ and shows that, in *base b*, the required number is precisely equal to

*the smallest power k for which $b^k$ occurs in column 1.*

6. Let $p$ be a prime number and imagine the whole numbers to be arranged in $p$ columns as before.

(i) Show that

*if p divides $b^p - b$ exactly, then either b lies in column 0 or $b^{p-1}$ lies in column 1.*

(ii) Show, by induction on $b$, that

$$p \text{ divides } b^p - b \text{ for every positive whole number } b.$$

[Hint: In the induction step from $b$ to $b + 1$, expand

$$(b + 1)^p - (b + 1)$$

using the binomial theorem. Then show that

*if $0 < i < p$, then p divides the binomial coefficient $\binom{p}{i}$ exactly.*]

(iii) Use part (ii) to show that, if $b$ is not a multiple of $p$, then $b^{p-1}$ lies in column 1. Conclude that if $k$ is the smallest power for which $b^k$ occurs in column 1, then $k$ divides $p - 1$.

7. Show that for any fraction $a/b$ it is possible to find whole numbers $c$, $m$, $k$ such that

$$\frac{a}{b} = \frac{c}{10^m(10^k - 1)}.$$

# The Fundamental Property of Real Numbers

Order is what is needed: all the thoughts that
can come into the human mind must be arranged
in an order like the natural order of numbers.
Descartes to Mersenne, 20 November 1629

Suppose I write down the first two infinite decimals that come into my head:

$$1\ 2\ 3 \cdot 4\ 5\ 6\ 7\ 8\ 9\ 6\ 7\ 8\ 9\ 6\ 7\ 8\ 9\ 6\ 7\ 8\ 9\ 6\ 7 \ldots,$$

$$1\ 2\ 3 \cdot 4\ 5\ 6\ 7\ 8\ 9\ 1\ 0\ 1\ 1\ 1\ 2\ 1\ 3\ 1\ 4\ 1\ 5\ 1\ 6 \ldots.$$

*How can I be sure that they automatically correspond to genuine numbers?*
If that sounds like a silly question, perhaps you need to be reminded that, strictly speaking, an *endless* string of digits remains meaningless until we find some convincing procedure whereby its endlessness can be interpreted in a usable way. *In the case of recurring decimals* we have gone to great lengths to explain as carefully as we can how such an endless string of digits can be said to represent a number: in fact we showed that the recurring decimals correspond precisely to the most familiar kind of numbers—the fractions. In particular the discussion in Chapter II.7 shows that

$$123.45\overset{\cdot}{6}78\overset{\cdot}{9} = \frac{12345}{10^2} + \frac{6789}{10^6(10^4 - 1)}.$$

But what about the second of the two infinite decimals at the beginning of this section:

$$1\ 2\ 3 \cdot 4\ 5\ 6\ 7\ 8\ 9\ 1\ 0\ 1\ 1\ 1\ 2\ 1\ 3\ 1\ 4\ 1\ 5\ 1\ 6\ 1\ 7 \cdots ?$$

A little thought is sufficient to convince oneself that this is *not* a recurring decimal, and so cannot be interpreted as easily as the first of our two infinite decimals. In particular, it cannot correspond to a fraction; so whatever *genuine number* it may happen to represent, it is unlikely to be one that we would recognise. But let us at least begin the line of attack which worked so well for recurring decimals.

Our first step was always to *interpret the endless decimal as an endless sum*: if we do the same in this case we obtain the endless sum

$$1 \cdot 10^2 + 2 \cdot 10 + 3 + 4 \cdot \frac{1}{10} + 5 \cdot \frac{1}{10^2} + 6 \cdot \frac{1}{10^3}$$

$$+ 7 \cdot \frac{1}{10^4} + 8 \cdot \frac{1}{10^5} + 9 \cdot \frac{1}{10^6} + 1 \cdot \frac{1}{10^7} + 0 \cdot \frac{1}{10^8} + \cdots .$$

Our second step was always to *interpret this endless sum in terms of the corresponding endless sequence of finite sums*: if we do the same in this case we obtain

$$1 \cdot 10^2, \ 1 \cdot 10^2 + 2 \cdot 10, \ 1 \cdot 10^2 + 2 \cdot 10 + 3,$$

$$1 \cdot 10^2 + 2 \cdot 10 + 3 + 4 \cdot \frac{1}{10}, \ 1 \cdot 10^2 + 2 \cdot 10 + 3 + 4 \cdot \frac{1}{10} + 5 \cdot \frac{1}{10^2}, \ \cdots$$

that is

$$100, \quad 120, \quad 123, \quad 123.4, \quad 123.45, \ldots .$$

Our third and final step was always *to find that number which appears to be the end point towards which this endless sequence of finite sums is heading, and to assign this numerical value to the original endless sum*. At this point we apparently come down to earth with a bump, for it is precisely the existence of such a number which is in doubt! However, if such a number were to exist, it would presumably be not only greater than each of the *finite* sums

$$100, \quad 120, \quad 123, \quad 123.4, \quad 123.45, \quad 123.456, \quad 123.4567, \ldots$$

but also less than each of the following

$$200, \quad 130, \quad 124, \quad 123.5, \quad 123.46, \quad 123.457, \quad 123.4568, \ldots .$$

There is no easy way out of our dilemma. Indeed, though it may look like begging the question, we have no choice but to appeal to a general fact about real numbers which we are in no position to prove. For it transpires that *perhaps the most important and fundamental property of the real numbers* is that

> *to any endless sequence of real numbers which keep increasing but which are all less than or equal to some number K, there always corresponds a real number $a \leq K$ towards which the endless sequence "appears to be heading."*

In our particular example, the endless sequence of finite sums

$$100, \quad 120, \quad 123, \quad 123.4, \quad 123.45, \quad 123.456, \ldots$$

certainly keeps increasing; and each of these finite sums is clearly less than 200 so we may choose $K = 200$ (though we could also choose $K = 130$, or

$K = 124$, or $K = 123.5$, etc.). So in this instance, the fundamental property of real numbers (stated above) assures us that there automatically exists a real number $a \leq K$ towards which our increasing sequence

$$100, \quad 120, \quad 123, \quad 123.4, \quad 123.45, \quad 123.456, \ldots$$

is *heading*. The difference between $a$ and the $n^{\text{th}}$ term of this sequence shrinks rapidly towards zero as $n$ increases, since

$$a - 100 < 10^2, \quad a - 120 < 10, \quad a - 123 < 1,$$

$$a - 123.4 < \frac{1}{10}, \quad a - 123.45 < \frac{1}{10^2}, \ldots,$$

—where we have chosen $K = 200$ to obtain the first inequality, $K = 130$ for the second, $K = 124$ for the third, $K = 123.5$ for the fourth, etc..

A similar argument can be used to conclude that *every infinite decimal automatically corresponds to a real number*. Moreover one can show (exactly as in Exercise 2 of Chapter II.8) that, with the single exception of the decimal fractions, distinct decimals correspond to distinct real numbers. It is also true that *every real number can be expressed as an infinite decimal*, though we cannot hope to prove either this fact or the fundamental property of real numbers until we make it absolutely clear what we mean by

*real numbers.*

Our failure to do just this is one of the main reasons why this section is so much more superficial than one would ideally like it to be—but though there are several different ways of giving the real numbers a strict and usable mathematical meaning, each approach involves subtleties which would be out of place at this stage[1].

At school the whole question is generally left in a very confused state. On the one hand the attempt to solve quadratic equations by means of the usual formula gives rise to specific numbers like $\sqrt{2}$, $\sqrt{3}$, $\sqrt{13}$, which do not seem to correspond to familiar fractions. But our continual use of square root tables and calculators encourages us to believe that such numbers can safely be written as finite decimals—and though we may be *told* that these numbers are not in fact *fractions*, this does not stop us thinking of them as though they were ordinary *finite decimals*. On the other hand we become vaguely aware of *endless decimals*, though probably without ever

---

[1] If this assertion merely strengthens your natural curiosity, then you should read Chapter 1 of *A Course in Pure Mathematics*, by G. H. Hardy (Cambridge University Press), or the more detailed and rather different approach in the first eight chapters—but especially Chapters 1, 2 and 8—of *Calculus* by M. Spivak (1st edition published by W. A. Benjamin/Addison Wesley, 1967; 2nd revised edition published by Publish or Perish Inc., 1980). After reading Chapters 1 and 8 of Spivak's book, you might work through Spivak's construction of the real numbers in Chapters 27 and 28.

doing anything with them[2]. And in the study of functions and the calculus we come increasingly to think of the real numbers as sitting, like points, on the axes of a graph, or on the so-called real number line. Now those who introduce *negative* numbers usually go to great lengths to relate their various aspects (for example: $-3$ is a new number which solves the previously insoluble problem *5 take away 8*; $-3$ is intimately related to the familiar number 3 and to the additon and subtraction operations via the identity $a + (-3) = a - 3$; and $-3$ also corresponds to a point on the number line). But our early experience of *real* numbers remains a tangled skein of *apparently distinct* threads:

(i) *inventing* symbols like $\sqrt{2}$ to produce solutions to previously insoluble equations like $x^2 - 2 = 0$;
(ii) *inventing arbitrary* endless decimals simply because some fractions seem to correspond to endless *recurring* decimals;
(iii) labelling points on an ordinary straight line.

And though we use the same real number language for all these threads, we spend very little time investigating the precise relationship between them.

Now whereas the beginner is naturally grateful for any connections whatever without worrying too much about their strictly logical justification, those who wish to put the calculus on a sound mathematical footing have no choice but to explain their every assumption and every deduction as carefully as they can. Perhaps the most surprising aspect of the 1870 cleaned-up version of the calculus is its implicit *assumption* that if one is hell-bent on describing the *essential logical structure* of the calculus, then *geometric real numbers* and *graphs* are more trouble than they are worth.[3]

In Part III we shall give examples of the kind of trouble which is in store for anyone who wishes to develop the calculus geometrically rather than arithmetically.

EXERCISES

1. (i) Let

$$a_n = \left(1 + \frac{1}{n}\right)^n.$$

Work out the first ten or so terms

$$a_1, a_2, a_3, a_4, a_5, a_6, a_7, a_8, a_9, a_{10}, \ldots .$$

---

[2] For example, it seems to be almost unheard of for endless recurring decimals to be examined in the light of what we learn about GP's. and geometric series.

[3] Though *geometric real numbers* and *graphs* are thereby excluded from the *official language of analysis*, they do not in practice get totally banished—merely "consigned to the margin." The potential motivating power of *geometrical* ideas—such as continuity, slope and area—is still valuable in helping us formulate those *purely arithmetic* notions, on the basis of which a strictly logical calculus can be developed.

What appears to be true about the terms of this sequence? Test your guess by working out the next few terms.

(ii) Show that if $i > 0$, then

$$\frac{n-i}{n} < \frac{n+1-i}{n+1}.$$

Deduce that if $r > 1$, then

$$\frac{n \cdot (n-1) \cdot \ldots \cdot (n-r+1)}{r!} \left(\frac{1}{n}\right)^r < \frac{(n+1) \cdot n \cdot \ldots \cdot (n-r+2)}{r!} \left(\frac{1}{n+1}\right)^r.$$

Use this to prove that if $n \geq 1$, then

$$\left(1 + \frac{1}{n}\right)^n < \left(1 + \frac{1}{n+1}\right)^{n+1}.$$

We have therefore *proved* that the terms of the endless sequence

$$a_1, \; a_2, \; a_3, \; a_4, \; a_5, \; a_6, \; \cdots$$

keep on increasing. If we can find a value of $K$ such that

$$a_i < K \quad \text{for each} \quad i \geq 1,$$

then it will follow from the fundamental property of real numbers that there must exist a real number $e \leq K$ towards which the endless sequence is heading.

2. (i) Expand

$$\left(1 + \frac{1}{n}\right)^n$$

by the binomial theorem. Show that if $n \geqslant 2$, then

$$\frac{n \cdot (n-1) \cdot \ldots \cdot (n-r+1)}{r!} \cdot \left(\frac{1}{n}\right)^r < \frac{1}{r!}$$

and conclude that if $n \geq 2$, then

$$\left(1 + \frac{1}{n}\right)^n < 1 + 1 + \frac{1}{2!} + \frac{1}{3!} + \cdots + \frac{1}{n!}.$$

Hence prove that for each $n \geq 1$

$$a_n = \left(1 + \frac{1}{n}\right)^n < 3.$$

(ii) Use a calculator to work out

$$a_{10}, \quad a_{100}, \quad a_{1000}, \quad a_{10000}, \quad a_{100000}, \quad \cdots$$

and so obtain as good an idea as possible of the actual value of the real number $e \leq 3$ towards which the endless sequence

$$a_1, \quad a_2, \quad a_3, \quad a_4, \quad a_5, \quad a_6, \quad \cdots$$

is heading.

# The Arithmetic of Infinite Decimals

To end our protracted encounter with infinite decimals we should at least answer the question which started it all off:

*Given that the familiar arithmetical procedures for addition, subtraction, multiplication and division simply do not work for infinite decimals, how can we possibly calculate*

$$\alpha + \beta, \quad \alpha - \beta, \quad \alpha \cdot \beta, \quad \frac{\alpha}{\beta}$$

*where $\alpha$ and $\beta$ are real numbers given in the form of infinite decimals?*

In answering this question we shall work with our original examples (page 67):

$$\alpha = .1 \; 2 \; 3 \; 4 \; 5 \; 6 \; 7 \; 8 \; 9 \; 1 \; 0 \; 1 \; 1 \; 1 \; 2 \; 1 \; 3 \; 1 \; 4 \; 1 \; 5 \; 1 \; 6 \; 1 \; 7 \; \ldots$$

$$\beta = .9 \; 8 \; 9 \; 9 \; 8 \; 9 \; 9 \; 9 \; 8 \; 9 \; 9 \; 9 \; 9 \; 8 \; 9 \; 9 \; 9 \; 9 \; 9 \; 8 \; 9 \; 9 \; 9 \; 9 \; 9 \; \ldots.$$

Recall that we have used the *fundamental property of real numbers* to identify $\alpha$ as that real number towards which the endless sequence

$$.1, \quad .12, \quad .123, \quad .1234, \quad .12345, \quad .123456, \ldots$$

is heading. Since each term of this sequence is less than .2, the fundamental property states that $\alpha \le .2$; in other words

$$\alpha - .1 \le .1$$

Since each term is also less than .13, the fundamental property states that $\alpha \le .13$; in other words

$$\alpha - .12 \le .01.$$

Thus, by choosing $K = .2$, $K = .13$, $K = .124$, etc., in turn, we obtain an *estimate* of the difference between $\alpha$ and each term of the endless sequence $.1, .12, .123, .1234, \ldots$ :

$$\alpha - .1 \le .1, \quad \alpha - .12 \le .01, \quad \alpha - .123 \le .001,$$

$$\alpha - .1234 \le .0001, \ldots.$$

If we introduce the symbols $a_1, a_2, a_3, \ldots$ to denote the terms of the endless sequence

$$a_1 = .1, \quad a_2 = .12, \quad a_3 = .123, \quad a_4 = .1234, \quad a_5 = .12345, \ldots,$$

then all these inequalities can be summarised in the form

$$(0 \leq)\alpha - a_i \leq \frac{1}{10^i}.$$

In the same way, we have used the fundamental property of real numbers to identify $\beta$ as that real number towards which the endless sequence

.9,   .98,   .989,   .9899,   .98998,   .989989,   .9899899, ...

is heading. If we introduce the symbols $b_1, b_2, b_3, \ldots$ to denote the terms of this endless sequence:

$$b_1 = .9,\quad b_2 = .98,\quad b_3 = .989,\quad b_4 = .9899,\quad b_5 = .98998, \ldots,$$

then we obtain inequalities estimating the difference between $\beta$ and $b_i$:

$$(0 \leq)\beta - b_i \leq \frac{1}{10^i}.$$

From this point on all we need to do is to follow our noses while keeping our wits about us. Since we have one increasing sequence

$$a_1,\quad a_2,\quad a_3,\quad a_4, \ldots$$

that *pinpoints* the real number $\alpha$, and another increasing sequence

$$b_1,\quad b_2,\quad b_3,\quad b_4, \ldots$$

that *pinpoints* the real number $\beta$, it seems not unreasonable to hope that *the "sum" of these two endless sequences*

$$a_1 + b_1,\quad a_2 + b_2,\quad a_3 + b_3,\quad a_4 + b_4, \ldots$$

*will be an increasing sequence that pinpoints $\alpha + \beta$ in the same way.*

Well, does it? Since the $a$'s and the $b$'s keep increasing, we know that, for each $i \geq 1$,

$$a_i \leq a_{i+1} \quad \text{and} \quad b_i \leq b_{i+1}.$$

From this it clearly follows that, for each $i \geq 1$,

$$a_i + b_i \leq a_{i+1} + b_{i+1};$$

hence the terms of our *"sum sequence"* certainly keep increasing. Also the $a$'s are all $\leq .2$, and the $b$'s are all $\leq 1.0$, so we can be sure that, for each $i \geq 1$,

$$a_i + b_i \leq 1.2.$$

But now the fundamental property of real numbers assures us that our *"sum sequence"* is necessarily heading for *some* real number, and we only have to explain

(i)  why this real number has to be $\alpha + \beta$, and
(ii) how we can calculate its infinite decimal.

At this point our inequalities

$$(0\leq)\alpha - a_i \leq \frac{1}{10^i} \quad \text{and} \quad (0\leq)\beta - b_i \leq \frac{1}{10^i}$$

join forces to give

$$(0\leq)\alpha + \beta - (a_i + b_i) \leq \frac{2}{10^i}.$$

In other words

(i) the term $a_i + b_i$ of our *"sum sequence"* differs from $\alpha + \beta$ by an amount $\leq 2/10^i$, *which shrinks rapidly towards zero as i increases.*

So the *"sum sequence"* does indeed pinpoint the real number $\alpha + \beta$ as we had hoped it would. Moreover, our inequalities pinpoint $\alpha + \beta$ not only in theory but also in practice. If we substitute the actual values of the $a$'s and $b$'s in the sum sequence, then we get

.1 + .9,   .12 + .98,   .123 + .989,   .1234 + .9899,   .12345 + .98998, ...

that is

1.0,   1.1,   1.112,   1.1133,   1.11343, ...;

if we now apply the inequality

$$0 \leq \alpha + \beta - (a_i + b_i) \leq \frac{2}{10^i}$$

for $i = 1, 2, 3, \ldots$ in turn, then we get

$$(a_1 + b_1 = )1.0 \leq \alpha + \beta \leq 1.0 + \left(\frac{2}{10}\right)$$

$$(a_2 + b_2 = )1.10 \leq \alpha + \beta \leq 1.10 + \left(\frac{2}{10^2}\right)$$

$$(a_3 + b_3 = )1.112 \leq \alpha + \beta \leq 1.112 + \left(\frac{2}{10^3}\right)$$

$$(a_4 + b_4 = )1.1133 \leq \alpha + \beta \leq 1.1133 + \left(\frac{2}{10^4}\right)$$

Etc.

Hence

$a_2 + b_2 = 1.10$ gives us the *first* decimal digit of $\alpha + \beta$ *exactly*

$a_3 + b_3 = 1.112$ gives us the first *two* decimal digits of $\alpha + \beta$ *exactly*

$a_4 + b_4 = 1.1133$ gives us the first *three* decimal digits of $\alpha + \beta$ *exactly*

Etc.

This provides us with a straightforward way of calculating *exactly*, for each $i \geq 1$, the first $i - 1$ digits in the infinite decimal for $\alpha + \beta$:

(ii) If the $i^{th}$ decimal digit of $a_i + b_i$ is *not a 9 or an 8* then the first $i - 1$ decimal digits of $\alpha + \beta$ are exactly the same as the first $i - 1$ decimal digits of $a_i + b_i$.
(If the $i^{th}$ decimal digit of $a_i + b_i$ is equal to 9 or 8 then we have to work slightly harder—see Exercise 1 parts (iii) and (iv).)

Before reading any further you should actually use this method to tackle Exercise 1.

*     *     *     *     *

We shall now deal as quickly as we can with the three remaining arithmetical operations, working all the time with the same two numbers $\alpha$, $\beta$ as before. It turns out that *subtraction* and *division* both give rise to a slightly unexpected complication,[1] so we shall look first at *multiplication*.

Remembering how effective the "sum" of our two increasing sequences

$$a_1, \quad a_2, \quad a_3, \quad a_4, \ldots$$
$$b_1, \quad b_2, \quad b_3, \quad b_4, \ldots$$

was in pinpointing $\alpha + \beta$, it is natural to hope that their "product"

$$a_1 b_1, \quad a_2 b_2, \quad a_3 b_3, \quad a_4 b_4, \ldots$$

will pinpoint $\alpha \cdot \beta$ in the same way. *Since the a's and b's are all positive*, (see Exercise 2), we can combine the inequalities

$$a_i \leq a_{i+1} \quad \text{and} \quad b_i \leq b_{i+1}$$

to obtain

$$a_i \cdot b_i \leq a_{i+1} \cdot b_{i+1};$$

hence the terms of our "product sequence" certainly keep increasing, and since the a's and b's are all less than .2 and 1.0 respectively, we can be sure that, for each $i \geq 1$,

$$a_i b_i \leq .2 \times 1.0 = .2.$$

Hence the fundamental property of real numbers assures us that the "product sequence" is necessarily heading for *some* real number. If we now make use of our inequalities

$$(0 \leq) \alpha - a_i \leq \frac{1}{10^i} \quad \text{and} \quad (0 \leq) \beta - b_i \leq \frac{1}{10^i}$$

---

[1] Perhaps this should not surprise us. After all the operations of *"subtracting a"* $(y \mapsto y - a)$ and of *"dividing by a"* $(y \mapsto (y/a))$ are defined as operations inverse to *"adding a"* and *"multiplying by a"*; and though inverse processes are easy to define, they quite often give rise to extra complications—square roots are more awkward than squares, and integration is more difficult than differentiation.

and the simple algebraic identity

$$\alpha \cdot \beta - a_i \cdot b_i = (\alpha - a_i) \cdot \beta + a_i \cdot (\beta - b_i)$$

then we get

$$(0 \leq) \alpha \cdot \beta - a_i \cdot b_i = (\alpha - a_i) \cdot \beta + a_i \cdot (\beta - b_i)$$

$$\leq \left(\frac{1}{10^i}\right) \cdot \beta + a_i \cdot \left(\frac{1}{10^i}\right)$$

$$= (\beta + a_i) \cdot \left(\frac{1}{10^i}\right)$$

and since $\beta \leq 1.0$ and $a_i \leq .2$ for each $i$, this means that

$$(0 \leq) \alpha \cdot \beta - a_i \cdot b_i \leq \left(\frac{1.2}{10^i}\right).$$

In other words, the term $a_i \cdot b_i$ of our "product sequence" differs from $\alpha \cdot \beta$ by an amount $\leq (1.2/10^i)$, *which shrinks rapidly towards zero as i increases.* Hence the "product sequence" not only pinpoints $\alpha \cdot \beta$, but does so in such a way that

$$(a_1 \cdot b_1 =).09 \leq \alpha \cdot \beta \leq .09 + \left(\frac{1.2}{10}\right)$$

$$(a_2 \cdot b_2 =).1176 \leq \alpha \cdot \beta \leq .1176 + \left(\frac{1.2}{10^2}\right)$$

$$(a_3 \cdot b_3 =).121647 \leq \alpha \cdot \beta \leq .121647 + \left(\frac{1.2}{10^3}\right)$$

$$(a_4 \cdot b_4 =).12215366 \leq \alpha \cdot \beta \leq .12215366 + \left(\frac{1.2}{10^4}\right)$$

Etc.

We can therefore be sure that

$a_2 \cdot b_2$ gives us the *first* decimal digit of $\alpha \cdot \beta$ *exactly*
$a_3 \cdot b_3$ gives us the first *two* decimal digits of $\alpha \cdot \beta$ *exactly*
$a_4 \cdot b_4$ gives us the first *three*[2] decimal digits of $\alpha \cdot \beta$ *exactly*
Etc.

---

[2] If the $i^{th}$ decimal digit of $a_i \cdot b_i$ is an 8 or a 9, then $a_i \cdot b_i$ might not give us the first $i - 1$ decimal digits of $\alpha \cdot \beta$ exactly. We must also *resist* any temptation to infer anything at all from the extra digits in, for example,

$$a_4 \cdot b_4 = .12215366:$$

their appearance of great precision is entirely spurious.

Before reading any further you should use this method to tackle Exercise 4.

$$* \quad * \quad * \quad * \quad *$$

In seeking to pinpoint $\beta - \alpha^3$ and $\alpha/\beta$ we now almost automatically write down the "difference sequence"

$$b_1 - a_1, \quad b_2 - a_2, \quad b_3 - a_3, \quad b_4 - a_4, \quad b_5 - a_5, \dots$$

and the "quotient sequence"

$$\frac{a_1}{b_1}, \quad \frac{a_2}{b_2}, \quad \frac{a_3}{b_3}, \quad \frac{a_4}{b_4}, \quad \frac{a_5}{b_5}, \dots$$

in the hope that our inequalities

$$(0 \le )\alpha - a_i \le \frac{1}{10^i} \quad \text{and} \quad (0 \le )\beta - b_i \le \frac{1}{10^i}$$

can be used to show that the two sequences we have written down really do pinpoint $\beta - \alpha$ and $\alpha/\beta$ respectively.

Consider the "difference sequence" first. Although we can write

$$(\beta - \alpha) - (b_i - a_i) = (\beta - b_i) - (\alpha - a_i),$$

you should be aware of the fact that *inequalities* cannot be simply "subtracted" (see Exercise 2(iii)). Instead, the inequalities for $\alpha - a_i$ and $\beta - b_i$ only allow us to conclude that

$$-\left(\frac{1}{10^i}\right) \le (\beta - b_i) - (\alpha - a_i) \le \left(\frac{1}{10^i}\right);$$

that is,

$$-\left(\frac{1}{10^i}\right) \le (\beta - \alpha) - (b_i - a_i) \le \left(\frac{1}{10^i}\right).$$

Now just in case we had not realised right at the start, this inequality suggests rather clearly that we must expect some of the terms $b_i - a_i$ to be less than $\beta - \alpha$, while others may be greater than $\beta - \alpha$. We cannot therefore expect the terms of the "difference sequence" to keep increasing—and in our particular example we find

$$b_1 - a_1 \le b_2 - a_2 \le \dots \le b_8 - a_8 > b_9 - a_9 \le b_{10} - a_{10} \le \dots$$

(see Exercise 5). This means that we cannot appeal to the fundamental

---

[3] For the sake of convenience we shall work with the positive difference $\beta - \alpha$ rather than $\alpha - \beta$.

property of real numbers to confirm that our "difference sequence" is definitely heading towards *some* real number. But we have now had so much practice with inequalities that we can see directly from

$$-\left(\frac{1}{10^i}\right) \le (\beta - \alpha) - (b_i - a_i) \le \left(\frac{1}{10^i}\right)$$

that the *absolute value* of the amount by which $b_i - a_i$ differs from $\beta - \alpha$ is always $\le (1/10^i)$, and so clearly *shrinks rapidly towards zero as i increases*. Thus our "difference sequence" does indeed pinpoint $\beta - \alpha$, and does so in such a way that

> *if the $i^{th}$ decimal place of $b_i - a_i$ is neither a 0 nor a 9, then $b_i - a_i$ definitely gives us the first $i - 1$ decimal digits of $\beta - \alpha$ exactly.*

The reader should use this method to calculate $\beta - \alpha$ (Exercise 6).

Finally the "quotient sequence"

$$\frac{a_1}{b_1}, \quad \frac{a_2}{b_2}, \quad \frac{a_3}{b_3}, \quad \frac{a_4}{b_4}, \quad \frac{a_5}{b_5}, \quad \dots$$

has to be treated in a similar way. This time the algebra is more unpleasant, but the principle is exactly the same. First we try to express the difference

$$\frac{\alpha}{\beta} - \frac{a_i}{b_i}$$

in a way which will allow us to feed in the inequalities

$$(0 \le )\alpha - a_i \le \frac{1}{10^i} \quad \text{and} \quad (0\le)\beta - b_i \le \frac{1}{10^i}.$$

A little thought produces

$$\frac{\alpha}{\beta} - \frac{a_i}{b_i} = \alpha\left(\frac{1}{\beta} - \frac{1}{b_i}\right) + \frac{1}{b_i}(\alpha - a_i)$$

$$= \frac{\alpha}{\beta \cdot b_i}(b_i - \beta) + \frac{1}{b_i}(\alpha - a_i).$$

Since the terms $(b_i - \beta)$ and $(\alpha - a_i)$ are of opposite sign we cannot expect that the terms $a_i/b_i$ *increase* towards $\alpha/\beta$ (see Exercise 7). Nevertheless, if we make use of the simple modulus inequality

$$|x + y| \le |x| + |y|,$$

then we obtain an estimate for the *absolute value* of the amount by which $a_i/b_i$ differs from $\alpha/\beta$:

$$\left| \frac{\alpha}{\beta} - \frac{a_i}{b_i} \right| = \left| \frac{\alpha}{\beta \cdot b_i} (b_i - \beta) + \frac{1}{b_i} (\alpha - a_i) \right|$$

$$\leq \left| \frac{\alpha}{\beta \cdot b_i}(b_i - \beta) \right| + \left| \frac{1}{b_i}(\alpha - a_i) \right|$$

$$= \frac{\alpha}{\beta \cdot b_i}(\beta - b_i) + \frac{1}{b_i}(\alpha - a_i)$$

$$\leq \frac{\alpha}{\beta \cdot b_i} \cdot \frac{1}{10^i} + \frac{1}{b_i} \cdot \frac{1}{10^i}$$

$$= \left( \frac{\alpha + \beta}{\beta b_i} \right) \cdot \frac{1}{10^i}.$$

But since we know that

$$\alpha \leq .2, \quad .9 \leq \beta \leq 1.0, \quad \text{and} \quad .9 \leq b_i \quad \text{(for each } i \geq 1),$$

we can replace the awkward-looking term by a larger constant:

$$\frac{\alpha + \beta}{\beta \cdot b_i} \leq \frac{.2 + 1.0}{(.9) \cdot (.9)} < \frac{3}{2}.$$

We therefore finish up with the simple statement that, for each $i \geq 1$,

$$\left| \frac{\alpha}{\beta} - \frac{a_i}{b_i} \right| < \frac{3}{2} \cdot \frac{1}{10^i}.$$

The reader should use this method to calculate $\alpha/\beta$ (Exercise 8).


### EXERCISES

*If you use a calculator at all for these Exercises, you will have to use it intelligently: the whole point of Exercises 1, 4, 6 and 8 is that we must know certain digits exactly, and in these four questions it would seem best to use a calculator only to check, and to assist, routine hand calculations.*

Throughout these Exercises $\alpha$, $\beta$ are the real numbers with infinite decimals

$$\alpha = .1\ 2\ 3\ 4\ 5\ 6\ 7\ 8\ 9\ 1\ 0\ 1\ 1\ 1\ 2\ 1\ 3\ 1\ 4\ 1\ 5\ 1\ 6\ 1\ 7\ldots$$

$$\beta = .9\ 8\ 9\ 9\ 8\ 9\ 9\ 9\ 8\ 9\ 9\ 9\ 9\ 8\ 9\ 9\ 9\ 9\ 9\ 8\ 9\ 9\ 9\ 9\ 9\ 9\ 8\ldots$$

and

$$a_1 = .1, \quad a_2 = .12, \quad a_3 = .123, \quad a_4 = .1234, \quad a_5 = .12345, \quad \text{etc.,}$$

$$b_1 = .9, \quad b_2 = .98, \quad b_3 = .989, \quad b_4 = .9899, \quad b_5 = 98998, \quad \text{etc..}$$

1. In the text we calculated the first three decimal places of $\alpha + \beta$ *exactly* and found that

$$\alpha + \beta = 1.113??? \ldots .$$

Use the method described in the text to calculate exactly

(i) the first *seven* decimal places of $\alpha + \beta$;
(ii) the first *nine* decimal places of $\alpha + \beta$;
(iii) the first *ten* decimal places of $\alpha + \beta$;
(iv) the first *eleven* decimal places of $\alpha + \beta$.

2. (i) Prove that if $a, b, c, d$ are all *positive* real numbers with

$$a \leq b \quad \text{and} \quad c \leq d,$$

then we can always conclude that

$$ac \leq bd.$$

(ii) Find specific values of $a, b, c, d$ for which

$$a \leq b \quad \text{and} \quad c \leq d$$

but

$$ac \not\leq bd.$$

(iii) Find positive values of $a, b, c, d$ for which

$$a \leq b \quad \text{and} \quad c \leq d$$

but

$$a - c \not\leq b - d.$$

(iv) Find positive values of $a, b, c, d$ for which

$$a \leq b \quad \text{and} \quad c \leq d$$

but

$$\frac{a}{c} \not\leq \frac{b}{d}.$$

3. (i) Work out

$$12 \times 9, \quad 123 \times 9, \quad 1234 \times 9, \quad 12345 \times 9, \quad 123456 \times 9,$$
$$1234567 \times 9, \quad 12345678 \times 9, \quad 123456789 \times 9.$$

Explain!

(ii) Work out

$$12 \times 8, \quad 123 \times 8, \quad 1234 \times 8, \quad 12345 \times 8, \quad 123456 \times 8,$$
$$1234567 \times 8, \quad 12345678 \times 8, \quad 123456789 \times 8.$$

Explain!

4. Use the method described in the text to calculate exactly

(i) the first *four* decimal places of $\alpha \cdot \beta$;

(ii) the first *six* decimal places of $\alpha \cdot \beta$;

(iii) the first *seven* decimal places of $\alpha \cdot \beta$;

(iv) the first *eight* decimal places of $\alpha \cdot \beta$.

5. (i) Work out the first *ten* terms of the "difference sequence"

$$b_1 - a_1, \quad b_2 - a_2, \quad b_3 - a_3, \quad b_4 - a_4, \quad b_5 - a_5, \ldots$$

exactly, and check the claim made on page 128 that

$$b_1 - a_1 \le b_2 - a_2 \le b_3 - a_3 \le \ldots \le b_8 - a_8 > b_9 - a_9 \le b_{10} - a_{10} \le \ldots.$$

Find *the next* value of $i > 8$ for which

$$b_i - a_i > b_{i+1} - a_{i+1}.$$

(ii) Use intelligent trial and error to invent two infinite decimals

$$\gamma = .x_1 x_2 x_3 x_4 \ldots, \quad \text{and} \quad \delta = .y_1 y_2 y_3 y_4 \ldots$$

for which the terms of the "difference sequence"

$$.x_1 - .y_1, \quad .x_1 x_2 - .y_1 y_2, \quad .x_1 x_2 x_3 - .y_1 y_2 y_3, \quad .x_1 x_2 x_3 x_4 - .y_1 y_2 y_3 y_4, \ldots$$

are alternately greater than and less than $\gamma - \delta$.

6. (i) Use the inequality

$$-\left(\frac{1}{10^i}\right) \le \beta - \alpha - (b_i - a_i) \le \left(\frac{1}{10^i}\right)$$

to complete the following and to extend the list as far as $b_{10} - a_{10}$:

$$b_1 - a_1 = .8 \qquad\qquad \text{so} \qquad .7 \le \beta - \alpha \le .9$$

$$b_2 - a_2 = \underline{\qquad} \qquad\qquad \text{so} \underline{\qquad} \le \beta - \alpha \le \underline{\qquad}$$

$$b_3 - a_3 = \underline{\qquad} \qquad\qquad \text{so} \underline{\qquad} \le \beta - \alpha \le \underline{\qquad}$$

(ii) Find the first *eight* decimal digits of $\beta - \alpha$ *exactly*.

(iii) Find the first *twelve* decimal digits of $\beta - \alpha$ *exactly*.

7. (i) Use a calculator to work out the first seven or eight terms of the "quotient sequence"

$$\frac{a_1}{b_1}, \quad \frac{a_2}{b_2}, \quad \frac{a_3}{b_3}, \quad \frac{a_4}{b_4}, \quad \frac{a_5}{b_5}, \ldots.$$

Why might you have expected the terms of this sequence to keep increasing at first? [Hint: Look carefully at the two terms on the right hand side of

$$\frac{\alpha}{\beta} - \frac{a_i}{b_i} = \frac{\alpha}{\beta \cdot b_i} (b_i - \beta) + \frac{1}{b_i} (\alpha - a_i).]$$

(ii) Show that when $i = 10$ we do in fact obtain

$$\frac{a_{i-1}}{b_{i-1}} < \frac{a_{i+1}}{b_{i+1}} < \frac{\alpha}{\beta} < \frac{a_i}{b_i}.$$

What is the *next* value of $i > 10$ for which

$$\frac{\alpha}{\beta} < \frac{a_i}{b_i} ?$$

(iii) Use intelligent trial and error (and a calculator) to invent two infinite decimals

$$\gamma = .x_1 x_2 x_3 x_4 \ldots \quad \text{and} \quad \delta = .y_1 y_2 y_3 y_4 \ldots$$

for which the terms of the quotient sequence

$$\frac{.x_1}{.y_1}, \quad \frac{.x_1 x_2}{.y_1 y_2}, \quad \frac{.x_1 x_2 x_3}{.y_1 y_2 y_3}, \quad \frac{.x_1 x_2 x_3 x_4}{.y_1 y_2 y_3 y_4}, \ldots$$

are alternately greater than and less than $\gamma/\delta$.

8. (i) Use the inequality

$$\left| \frac{\alpha}{\beta} - \frac{a_i}{b_i} \right| < \frac{3}{2} \cdot \frac{1}{10^i}$$

to complete the following and to extend the list as far as $a_8/b_8$:

$$\frac{a_1}{b_1} = .1 \qquad \text{so} \qquad -.038 < \frac{\alpha}{\beta} < .261$$

$$\frac{a_2}{b_2} = .12244 \ldots \quad \text{so} \quad .10744 \ldots < \frac{\alpha}{\beta} < .13744 \ldots$$

$$\frac{a_3}{b_3} = \underline{\quad\quad} \qquad \text{so} \underline{\quad\quad} < \frac{\alpha}{\beta} < \underline{\quad\quad}$$

(ii) Find the first *six* decimal digits of $\alpha/\beta$ exactly.
(iii) Find the first *eight* decimal digits of $\alpha/\beta$ exactly.

CHAPTER II.12

# Reflections on Recurring Themes

We began Part II by examining (in Chapters II.1, II.3 and II.5) the relationship between three ideas:

(1) common measures for pairs of line segments AB, CD in geometry;
(2) ordinary rational numbers $a/b$; and
(3) (highest) common factors of $a$ and $b$.

In Chapter II.3 we devised a procedure, based on the game of Euclid, which automatically produced the highest common factor $hcf(a, b)$ for any two whole numbers $a, b$. This procedure was then adapted to obtain a general method for finding the greatest common measure of any two segments AB, CD—assuming, that is, that they actually have a common measure. But then we applied this general procedure to the side and diagonal of a square, or a regular pentagon: and though it still *worked* in the sense that each step could be comfortably carried out, it quickly became clear that the *sequence of steps* was following an endlessly recurring pattern, and would therefore never actually produce a common measure. We were thus forced to the conclusion that *the side and diagonal of a square have no common measure.*

We then abandoned the specific problems of producing the highest common factor of a pair of whole numbers, and of finding the greatest common measure of a pair of line segments, in order to consider general properties of real numbers and arithmetic.

In Chapter II.5 we discussed the *endless counting process* as being, in some sense, the source of all infinite processes; and we went out of our way to stress that it is our use of a *particularly fortunate notation* which makes this astonishing idea so easy to accept. We considered afresh the familiar decimal notation for decimal fractions, and showed why it is quite correct to use the familiar division process to work out the decimal representation of a fraction such as $3/40 = .075$. But then in Chapter II.6 we applied the same procedure to the fraction $3/39 = 1/13$: and though it still *worked* in the sense that *each step* could be comfortably carried out, it quickly became clear that the *sequence of steps* was following an endlessly recurring pattern, and would therefore never actually produce an ordinary decimal. We were thus forced to the conclusion that *the fraction 3/39 could not be expressed as an ordinary decimal.*

But rather than stop there, as we did in the case of common measures, we took courage from the fact that the *endless decimal* $0.\dot{0}7692\dot{3}$ which emerged from our division procedure, appeared to capture something of the nature of 1/13: we therefore decided to dig a little deeper (—the fact that decimals feel more familiar no doubt also helped). In Chapters II.6 and II.7 we tried to give a meaning to the *endless decimal* which emerged from our division procedure by writing it first as an *endless sum*; this *endless* sum we then interpreted as meaning the "endpoint," towards which the *endless sequence* of finite sums was heading. The whole procedure worked beautifully for *recurring decimals.* since in that case we managed to find a fairly convincing way of actually calculating the "endpoint" towards which the endless sequence of finite sums was heading. And in Chapter II.9 we eventually showed that *recurring decimals correspond precisely to ordinary fractions.* But when we applied the same procedure to such endless decimals as

$$1\,2\,3\,.\,4\,5\,6\,7\,8\,9\,1\,0\,1\,1\,1\,2\,1\,3\,1\,4\,1\,5\,1\,\ldots$$

$$.\,1\,2\,3\,4\,5\,6\,7\,8\,9\,1\,0\,1\,1\,1\,2\,1\,3\,1\,4\,\ldots$$

and

$$.\,9\,8\,9\,9\,8\,9\,9\,9\,8\,9\,9\,9\,9\,8\,9\,9\,9\,9\,9\,\ldots,$$

though it still *worked* in the sense that one could comfortably write down the corresponding *endless sums,* and the *endless sequences* of finite sums, it quickly became clear that the *endlessness* in these cases does not arise in a manageable (recurring) form. We were thus forced to the conclusion that *such endless decimals could not possibly correspond to ordinary fractions;* moreover, *if they represented numbers at all, then we would probably require some entirely different way of identifying them as numbers.* It was this that led us to introduce the fundamental property of real numbers in Chapter II.10.

In Chapter II.5 we reviewed the usual arithmetical procedures of addition, subtraction, multiplication and division for ordinary decimals. Now most of us had presumably come to take the arithmetic of real numbers more or less for granted. But when we actually tried to devise similar arithmetical procedures for infinite decimals, it was not even clear how to begin. And though we eventually succeeded in Chapter II.11, our resolution of the problem of adding, subtracting, multiplying and dividing infinite decimals hinted strongly that in general, infinite decimals and the associated infinite processes have a character which our previous examples had concealed. For example, we managed to calculate the sum $\alpha + \beta$ of two arbitrary infinite decimals $\alpha$, $\beta$ in the following sense: to obtain the first decimal place of $\alpha + \beta$ we needed to know the first few decimal places of $\alpha$ and $\beta$

separately,[1] and to apply the fundamental property with one particular choice of the upper bound $K$; to obtain the second decimal place of $\alpha + \beta$, we needed to know $\alpha$ and $\beta$ a bit more accurately and to apply the fundamental property again with a new choice of $K$; to obtain the third decimal place of $\alpha + \beta$ we needed yet a third refinement of the upper bound $K$; and so on. Thus in one sense the numbers

$$\alpha = .1\ 2\ 3\ 4\ 5\ 6\ 7\ 8\ 9\ 1\ 0\ 1\ 1\ 1\ 2\ 1\ 3\ 1\ 4\ \ldots$$

$$\beta = .9\ 8\ 9\ 9\ 8\ 9\ 9\ 9\ 8\ 9\ 9\ 9\ 9\ 8\ 9\ 9\ 9\ 9\ 9\ \ldots$$

used in the text are misleading: for though these infinite decimals do not recur, we nevertheless feel that we really know *the whole infinite decimal for* $\alpha$ (and for $\beta$). In contrast the infinite decimal for $\alpha + \beta$ is only known *in principle*: that is, though $\alpha + \beta$ is uniquely determined, I can never really pretend to know *the whole infinite decimal for* $\alpha + \beta$; but if you specify a positive whole number $N$, then I can always calculate the first $N$ decimal places

$$\alpha + \beta = 1.a_1 a_2 a_3 \ldots a_N \ldots$$

*exactly.*

We shall leave this matter here, but you will almost certainly need to come back to it again.

In the final chapter of Part II we return to our original procedure for finding the greatest common measure of two segments AB, CD, and interpret it as a way of finding a *numerical value* for the ratio AB/CD. The form in which we obtain this numerical value may appear rather cumbersome, but it has proved its worth in many branches of mathematics, and—perhaps more important for us here—provides an elementary example of the inadequacy of plain intuition when dealing with infinite processes.

---

[1] We showed that $|\alpha + \beta - (a_i + b_i)| < 2/10^i$. This is usually enough to ensure that the first decimal place of $\alpha + \beta$ can be calculated from the first few decimal places of $\alpha$ and $\beta$. But this is by no means always the case, and one must be careful. For example, what is the first decimal place of

$$.123456789012345678901234\ldots$$
$$+\ .876543210987654321098765\ldots$$

# Continued Fractions

Recall the procedure we introduced in Chapter II.3 to find the greatest common measure of two line segments AB, CD. First we subtract CD from AB as many times as possible (Figure 30)

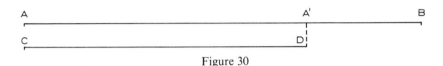

Figure 30

leaving A′B; then A′B < CD, and we turn our attention from the pair (AB, CD) to the pair (CD, A′B). Next subtract A′B from CD as many times as possible (Figure 31), leaving C′D;

Figure 31

then C′D < A′B, and we turn our attention from the pair (CD, A′B) to the pair (A′B, C′D). We continue in this way until we obtain a pair (UV, XY) where XY fits into UV a whole number of times *leaving no remainder*—in which case XY is the required greatest common measure of the original pair of line segments AB, CD.

This chapter is motivated by two questions:

(1) Since AB, CD have a common measure precisely when AB/CD is a fraction, can the above (finite) procedure be interpreted to give a (finite) numerical representation of the ratio AB/CD in this case?

(2) If the ratio AB/CD is irrational, then the procedure continues endlessly. *Can it nevertheless be interpreted to give a (presumably endless) numerical representation for the ratio AB/CD even when this ratio is irrational?*

We shall start by answering question (1), and will investigate very briefly the curious representation for fractions which results. Finally we shall return to consider question (2).

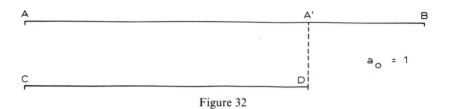

Figure 32

Suppose that CD fits into AB precisely $a_0$ times, leaving remainder A'B (Figure 32). Then

$$\frac{AB}{CD} = \frac{AA' + A'B}{CD} = \frac{AA'}{CD} + \frac{A'B}{CD} = a_0 + \frac{A'B}{CD}.$$

In particular, since A'B < CD, we know that

$$a_0 < \frac{AB}{CD} < a_0 + 1.$$

We then turn our attention away from the pair (AB, CD) and the corresponding ratio AB/CD to consider the pair (CD, A'B) and *the corresponding ratio CD/A'B*, and repeat exactly the same process. Suppose that A'B fits into CD precisely $a_1$ times leaving remainder C'D (Figure 33).

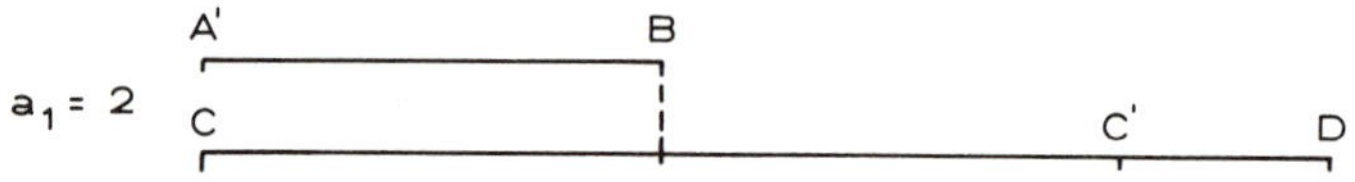

Figure 33

Then

$$\frac{CD}{A'B} = \frac{CC' + C'D}{A'B} = \frac{CC'}{A'B} + \frac{C'D}{A'B} = a_1 + \frac{C'D}{A'B}.$$

Observe that if we wish to substitute this back in our original expression for AB/CD then we must rewrite the original expression in terms of CD/A'B rather than A'B/CD: thus

$$\frac{AB}{CD} = a_0 + \frac{A'B}{CD} = a_0 + \frac{1}{CD/A'B}$$

$$= a_0 + \frac{1}{a_1 + \dfrac{C'D}{A'B}}$$

In particular, since C'D < A'B, we know (Exercise 1) that

$$a_0 < a_0 + \frac{1}{a_1 + 1} < \frac{AB}{CD} < a_0 + \frac{1}{a_1} \le a_0 + 1.$$

We then turn our attention away from the pair (CD, A'B) and the corresponding ratio CD/A′B to consider the pair (A′B, C′D) and *the corresponding ratio A′B/C′D*, and repeat exactly the same process. Suppose that C′D fits into A′B precisely $a_2$ times leaving remainder A″B (Figure 34).

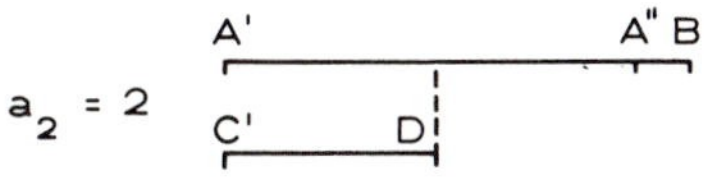

Figure 34

Then

$$\frac{A'B}{C'D} = \frac{A'A'' + A''B}{C'D} = \frac{A'A''}{C'D} + \frac{A''B}{C'D} = a_2 + \frac{A''B}{C'D}$$

and

$$\frac{AB}{CD} = a_0 + \cfrac{1}{a_1 + \cfrac{C'D}{A'B}} = a_0 + \cfrac{1}{a_1 + \cfrac{1}{A'B/C'D}} = a_0 + \cfrac{1}{a_1 + \cfrac{1}{a_2 + A''B/C'D}}$$

In particular, since A″B < C′D, we know (Exercise 2) that

$$a_0 < a_0 + \cfrac{1}{a_1 + 1} \le a_0 + \cfrac{1}{a_1 + \cfrac{1}{a_2}} < \frac{AB}{CD} < a_0 + \cfrac{1}{a_1 + \cfrac{1}{a_2 + 1}} < a_0 + \frac{1}{a_1} \le a_0 + 1$$

We continue in this way until we obtain a pair (UV, XY), with XY < UV, for which XY fits precisely $a_n$ times into UV *leaving no remainder* (Figure 35).

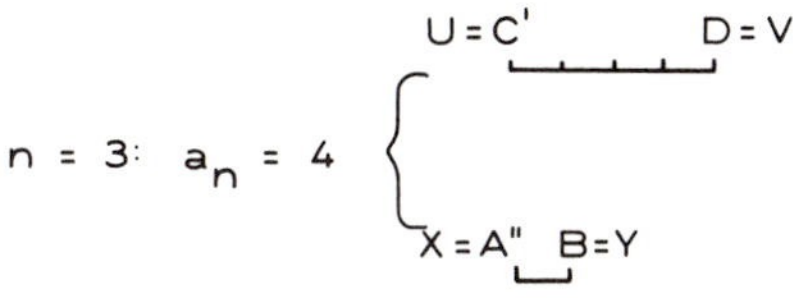

Figure 35

Then

$$\frac{UV}{XY} = a_n + 0$$

and

$$\frac{AB}{CD} = a_0 + \cfrac{1}{a_1 + \cfrac{1}{a_2 + \cfrac{1}{\cdots + \cfrac{1}{a_n + 0}}}} \qquad (2.13.1)$$

Observe that, whereas some of the earlier $a_i$'s may well be equal to 1, the last $a_n$ is always $\geq 2$ (since XY < UV).[1] Thus we obtain an expression which looks like (2.13.1) in which

   (i) $a_0, a_1, a_2, \ldots, a_n$ are whole numbers,
   (ii) $a_1, a_2, \ldots, a_n$ are positive,
  (iii) $a_0 \geq 0$ (and $a_0 = 0$ precisely when AB < CD),
  (iv) $a_n \geq 2$ (unless AB = CD, in which case $n = 0$, AB/CD = 1).

    Any number written in this form is called a *continued fraction*. Thus we obtain a clear, if slightly curious, answer to question (1): *if AB, CD have a common measure, then our procedure for finding the greatest common measure of AB, CD can be interpreted so as to express the ratio AB/CD as a continued fraction.*

    If we work with numbers $a$, $b$ in place of line segments AB, CD, then it is easy to see that any (positive) fraction $a/b$ can be written *in precisely one way* as a continued fraction: for example (see page 41),

$$\frac{289}{27} = \frac{270 + 19}{27} = 10 + \frac{19}{27} = 10 + \frac{1}{27/19}$$

and

$$\frac{27}{19} = \frac{19 + 8}{19} = 1 + \frac{8}{19} = 1 + \frac{1}{19/8}$$

and

$$\frac{19}{8} = \frac{16 + 3}{8} = 2 + \frac{3}{8} = 2 + \frac{1}{8/3}$$

and

$$\frac{8}{3} = \frac{6 + 2}{3} = 2 + \frac{2}{3} = 2 + \frac{1}{3/2}$$

and

$$\frac{3}{2} = \frac{2 + 1}{2} = 1 + \frac{1}{2}$$

so

$$\frac{289}{27} = 10 + \cfrac{1}{1 + \cfrac{1}{2 + \cfrac{1}{2 + \cfrac{1}{1 + \cfrac{1}{2}}}}} \qquad (2.13.2)$$

---

[1] Unless $n = 0$ and AB = CD, in which case the procedure stops at the very first step.

To save space we shall often abbreviate continued fractions such as this by listing only the $a_i$'s: thus we would normally write

$$\frac{289}{27} = [10; 1, 2, 2, 1, 2] \tag{2.13.2}$$

and

$$\frac{AB}{CD} = [a_0; a_1, a_2, \ldots, a_n] \tag{2.13.1}$$

This makes life easy for the printer but much harder for the reader, and you should, at least to begin with, always have pencil and paper handy to write out these abbreviated continued fractions in full.

In order to familiarise yourself with these rather strange expressions you are advised to work through Exercise 3 before reading the rest of this chapter.

*    *    *    *    *

In the first part of this chapter we have shown how our procedure for finding the greatest common measure of two line segments AB, CD (which is exactly the same as our procedure for finding the highest common factor of two whole numbers $a$, $b$) can be interpreted to express the ratio AB/CD (or the fraction $a/b$) in the form of a continued fraction.

Now in the continued fraction for $a/b$

$$a/b = [a_0; a_1, a_2, \ldots, a_n]$$

$a_0$, for example, is simply the number of times that $b$ goes into $a$ and so is not affected if we multiply both numerator and denominator of $a/b$ by a common factor; the same is true of all the other $a_i$'s (Exercise 4). Two equivalent fractions therefore give rise to one and the same continued fraction; thus each (positive) rational number corresponds to precisely one continued fraction. Conversely, each continued fraction $[a_0; a_1, a_2, \ldots, a_n]$ clearly corresponds to some rational number, *so we could, if we wished, work with continued fractions instead of the more familiar representation of rational numbers in the form $a/b$.*

You might reasonably ask why anyone would ever voluntarily *choose* to work with continued fractions. They are certainly very cumbersome; worse still they appear to be quite useless for tasks involving arithmetic. Their real value arises from the fact that they provide us with *really good approximations to irrational numbers.* In the past, whenever you have wanted a fraction approximating $\pi$ or $\sqrt{2}$ you have presumably taken the first few digits of their decimal expansions

$$\pi = 3.141592653\ldots$$
$$\sqrt{2} = 1.414213562\ldots.$$

But, however convenient this method may appear to be, it has the essential drawback that it can only give *decimal fraction approximations*. Thus if we take the first few digits of the decimal expansion we obtain the following approximations:

$$3 < \pi < 4$$

$$3.1 = 3\,\frac{1}{10} < \pi < 3\,\frac{2}{10} = 3.2$$

$$3.14 = 3\,\frac{14}{100} < \pi < 3\,\frac{15}{100} = 3.15$$

$$3.141 = 3\,\frac{141}{100} < \pi < 3\,\frac{142}{100} = 3.142$$

But if one allows arbitrary fractions, then it is nearly always possible to do very much better than this.[2] Continued fractions provide in some sense the best possible approximations. We shall see in a moment how irrational numbers can be represented by *endless continued fractions*; if we jump the gun slightly, then we can compare the first few decimal approximations to $\pi$ (given above) with the *dramatically accurate approximations* that arise from the continued fraction for $\pi$:

$$\pi = 3 + \cfrac{1}{7 + \cfrac{1}{15 + \cfrac{1}{1 + \cfrac{1}{292 + \cfrac{1}{\ddots}}}}}$$

namely (Exercise 3(iii))

$$3 < \pi < 3\,\frac{1}{7} = 3.\dot{1}4285\dot{7}$$

$$3.14150943\ldots = 3\,\frac{15}{106} < \pi < 3\,\frac{16}{113} = 3.141592920\ldots$$

$$3.141592653\ldots = 3\,\frac{4687}{33102} < \pi < \ldots .$$

The reader who would like to know more about continued fractions should consult the excellent and readable little book by H. Davenport: *The Higher Arithmetic* (published by Hutchinson, 1968), C.D. Olds: *Continued Fractions* (published by the Mathematical Association of America, 1963), or

---

[2] Suppose, for example, that you were to use *decimal fractions* to find a good approximation to the fraction 1/3. Then you would obtain the decimal fractions 3/10, 33/100, etc. instead of the obvious choice—namely 1/3 itself.

A. Ya. Khinchin: *Continued Fractions* (published by the University of Chicago Press, 1964). But we must move on to explain why we have bothered to consider them in Part II at all.

We have seen that if two line segments AB, CD have a common measure, then our procedure for finding their greatest common measure can also be used to express the ratio AB/CD as a continued fraction

$$\frac{AB}{CD} = [a_0; a_1, a_2, \ldots, a_n]$$

where

CD fits into AB precisely $a_0$ times leaving remainder A'B,
A'B fits into CD precisely $a_1$ times leaving remainder C'D,
C'D fits into A'B precisely $a_2$ times leaving remainder A''B,
and so on.

Since AB and CD have a common measure, the procedure necessarily stops: at which stage one segment fits precisely $a_n$ times into the other segment *leaving no remainder*.

Now in Chapter II.4 (pages 54 to 56) we showed that when our procedure for finding the greatest common measure of two line segments is applied to the diagonal AD and side BC of a regular pentagon ABCDE *it simply goes on for ever*. We concluded that AD and BC have no common measure: in particular we cannot hope to interpret our procedure to express the ratio AD/BC as an (ordinary, finite) continued fraction.

*But suppose we try* (Figure 36)!

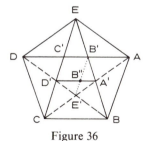

Figure 36

BC = B'D fits precisely *once* into AD leaving remainder AB' (so $a_0 = 1$).
AB' fits precisely *once* into AC' = BC leaving remainder B'C' (so $a_1 = 1$).
But now B'C' and AB' = A'D' are just the side and diagonal of the smaller pentagon A'B'C'D'E', so
B'C' fits precisely *once* into A'D' = AB' leaving remainder A'B'' (so $a_2 = 1$).
And so it goes on, with $a_i = 1$ at every stage.

It is therefore exceedingly tempting to write

$$\frac{AD}{BC} = [1; 1, 1, 1, 1, \ldots]$$

where the sequence of 1's is supposed to be *endless*.

To summarise:

(i) When two line segments have a common measure, we found a perfectly valid procedure which allows us to express their ratio as an ordinary, decent and perfectly respectable continued fraction.

(ii) We then applied the same procedure to the diagonal AD and side BC of a regular pentagon. And although the procedure as such is perfectly valid, we naively assumed (by analogy with the case where the two segments concerned have a common measure) that this allowed us to express the ratio AD/BC as an (*endless!*) *continued fraction*—although this assumption *could not possibly be justified in the same way*. All we can be sure of is that, when the procedure is expressed numerically, it generates the endless sequence

$$a_0 = 1, \quad a_1 = 1, \quad a_2 = 1, \quad a_3 = 1, \quad a_4 = 1, \ldots .$$

We may write this in the form

$$[1; 1, 1, 1, 1, \ldots]$$

if we please, but we have not as yet explained how any *numerical meaning* can be attributed to such tailless monsters.

*But suppose we try!* When we first set about the task of attributing some *clear mathematical meaning* to endless continued fractions we discover a twist similar to that which you probably found when trying to multiply

$$.9\ 8\ 9\ 9\ 8\ 9\ 9\ 9\ 8\ 9\ 9\ 9\ 9\ 8\ 9 \ldots \times .\dot{1}.$$

Multiplication of decimals presents the beginner with no problems as long as she/he can start with the right hand column and "*carry*" systematically. Similarly a beginner finds it easy (if slightly laborious) to simplify a continued fraction like

$$1 + \cfrac{1}{2 + \cfrac{1}{3 + \cfrac{1}{4}}}$$

though she/he naturally assumes that *one has to start at the bottom* (in this case with the 4) *and work up*. But for an *endless* continued fraction like

$$[1; 2, 3, 4, 5, 6, 7, 8, 9, 10, 11, 12, 13, 14, 15, 16, \ldots]$$

this approach is completely useless! Moreover, supposing for the moment that this tailless monster really does represent a number $\alpha$,[3] we have very

---

[3] $\alpha$ must presumably be *irrational* since each rational number gives rise to an ordinary (finite) continued fraction.

little feeling for the approximate size of $\alpha$, except that we presumably *expect* to find that

$$1 < 1 + \cfrac{1}{2 + \cfrac{1}{3 + 1 \atop \ddots}} < 1 + 1$$

But wait a minute! If we have *these* inequalities for $\alpha$, would we not also expect to find the next pair of inequalities

$$1 + \cfrac{1}{2 + 1} < \alpha < 1 + \frac{1}{2},$$

just as for ordinary continued fractions (see Exercises 1 and 2); and the next

$$1 + \cfrac{1}{2 + \cfrac{1}{3}} < \alpha < 1 + \cfrac{1}{2 + \cfrac{1}{3 + 1}} ;$$

and the next ... ?

Even though the symbol $\alpha$ has still not been given a meaning, these few wild thoughts point us firmly in the kind of direction we should have suspected from our encounter with endless decimals in Chapters II.6, II.7, and II.8. *Then*, we were in the happy position of understanding *ordinary (finite) decimals*, and so our first move was to interpret each *endless decimal* as an *endless sequence of ordinary (finite) decimals*. And even if we had temporarily forgotten all this, the above inequalities, which we suspect $\alpha$ must satisfy, suggest firmly that the number $\alpha$ which we are seeking will be sandwiched (alternately from below and above) between the terms of the endless sequence

$$[1], \quad [1; 2], \quad [1; 2, 3], \quad [1; 2, 3, 4], \quad [1; 2, 3, 4, 5], \ldots \quad (2.13.3)$$

of ordinary continued fractions.

We must therefore go right back to the beginning and interpret the (still meaningless!) *endless* continued fraction

$$[1; 2, 3, 4, 5, 6, 7, 8, 9, 10, 11, 12, 13, 14, 15, 16, \ldots]$$

in terms of the *endless sequence* (2.13.3) of ordinary finite continued fractions. Now each term of this endless sequence can be written as an ordinary fraction: if we do this, we obtain the endless sequence of ordinary fractions (Exercise 6):

$$1, \quad \frac{3}{2}, \quad \frac{10}{7}, \quad \frac{43}{30}, \quad \frac{225}{157}, \quad \frac{1393}{972}, \ldots \quad (2.13.3)$$

It is therefore natural to identify $\alpha$ as the (presumably irrational) real number *"towards which this endless sequence of fractions is heading."*

But how can we be sure that this endless sequence is heading anywhere at all? As always, we have two alternatives:

(i) We may try to find $\alpha$ *directly* as we did in Chapter II.7 (though this can only work if $\alpha$ is a sufficiently nice number which we somehow manage to identify explicitly);

(ii) alternatively we may appeal to the fundamental property of real numbers.

Since we have little hope of identifying the required $\alpha$ explicitly we are forced to use the second alternative. Our proof, that the endless sequence (2.13.3) is definitely heading towards a specific (if elusive) real number $\alpha$, has two steps.

*Step* 1. First (Exercise 7) we have to prove the (infinitely many) inequalities

$$[1] < [1; 2, 3] < [1; 2, 3, 4, 5] < \cdots < [1; 2, 3, 4, 5, 6] < [1; 2, 3, 4] < [1; 2]$$

We can then apply the fundamental property of real numbers (Chapter II.10) to the increasing sequence

$$[1] < [1; 2, 3] < [1; 2, 3, 4, 5] < [1; 2, 3, 4, 5, 6, 7] < \ldots,$$

each of whose terms is less than $K_1 = [1; 2]$ (or $K_1 = [1; 2, 3, 4]$, or $K_1 = [1; 2, 3, 4, 5, 6]$, or ...). This then ensures that the odd numbered terms in our original endless sequence are indeed heading towards some real number $\alpha_1 \leq K_1$.

We must then apply the fundamental property of real numbers to the *decreasing* sequence of even-numbered terms from our original endless sequence

$$[1; 2] > [1; 2, 3, 4] > [1; 2, 3, 4, 5, 6] > [1; 2, 3, 4, 5, 6, 7, 8] > \ldots$$

each of whose terms is greater than $K_2 = [1]$ (or $K_2 = [1; 2, 3]$, or $K_2 = [1; 2, 3, 4, 5]$, or ...). This then ensures (Exercise 8) that the even numbered terms in our endless sequence are also heading towards some real number $\alpha_2 \geq K_2$.

*Step* 2. It remains to show that $\alpha_1 = \alpha_2$. This we do by observing first that no matter which values one chooses for $K_1$ and $K_2$, they necessarily satisfy

$$K_2 \leq \alpha_1 \leq K_1 \quad \text{and} \quad K_2 \leq \alpha_2 \leq K_1.$$

Hence

$$|\alpha_2 - \alpha_1| \leq K_1 - K_2.$$

We therefore only need to show that, by choosing first

$$K_1 = [1; 2], \quad K_2 = [1]$$

then

$$K_1 = [1; 2, 3, 4], \quad K_2 = [1; 2, 3]$$

then

$$K_1 = [1; 2, 3, 4, 5, 6], \quad K_2 = [1; 2, 3, 4, 5]$$

and so on, we may make the positive number $K_1 - K_2$ as small as ever we please; it then necessarily follows that the *fixed* number $|\alpha_2 - \alpha_1|$ must be *smaller than all of these possible values for $K_1 - K_2$, and so has no choice but to be equal to zero!* Hence $\alpha_1 = \alpha_2$ and our endless sequence (2.13.3) does indeed define a single real number $\alpha (= \alpha_1 = \alpha_2)$. Assistance in writing out the details is offered in Exercise 9.

Thus if we interpret the *endless continued fraction*

$$[1; 2, 3, 4, 5, 6, 7, 8, 9, 10, 11, 12, 13, 14, 15, \ldots]$$

in terms of the *endless* sequence of ordinary continued fractions,

$$[1], \quad [1; 2], \quad [1; 2, 3], \quad [1; 2, 3, 4], \quad [1; 2, 3, 4, 5], \quad [1; 2, 3, 4, 5, 6], \ldots$$

then there is a definite real number $\alpha$ which is pinpointed by this endless sequence, and it is natural to interpret the original endless continued fraction as representing this number $\alpha$: we therefore write

$$\alpha = [1; 2, 3, 4, 5, 6, 7, 8, 9, 10, 11, 12, 13, 14, \ldots].$$

Now although the above discussion was presented in the context of one particular endless continued fraction, the two crucial steps (Exercises 7 and 9) are *completely general*. If we start with *any* endless continued fraction

$$[a_0; a_1, a_2, a_3, a_4, a_5, a_6, \ldots]$$

and interpret it in terms of the corresponding endless sequence of ordinary continued fractions

$$[a_0], \quad [a_0; a_1], \quad [a_0; a_1, a_2], \quad [a_0; a_1, a_2, a_3], \quad [a_0; a_1, a_2, a_3, a_4], \ldots$$

then, by Exercise 7, we always have

$$[a_0] < [a_0; a_1, a_2] < [a_0; a_1, a_2, a_3, a_4] < \ldots$$
$$< [a_0; a_1, a_2, a_3, a_4, a_5] < [a_0; a_1, a_2, a_3] < [a_0, a_1].$$

We can then apply the fundamental property of real numbers to the increasing sequence

$$[a_0] < [a_0; a_1, a_2] < [a_0; a_1, a_2, a_3, a_4] < \ldots$$

each of whose terms is less than $K_1 = [a_0; a_1]$ (or $K_1 = [a_0; a_1, a_2, a_3]$, or $\ldots$). This then ensures that the odd numbered terms in our original endless sequence are heading towards some real number $\alpha_1 \leq K_1$. We must then apply the fundamental property of real numbers to the decreasing sequence of even-numbered terms from our original endless sequence:

$$[a_0; a_1] > [a_0; a_1, a_2, a_3] > [a_0; a_1, a_2, a_3, a_4, a_5] > \ldots$$

each of whose terms is greater than $K_2 = [a_0]$ (or $K_2 = [a_0; a_1, a_2]$, or $\ldots$). This then ensures that the even numbered terms are also heading towards some real number $\alpha_2 \geq K_2$. We can then show that $\alpha_1 = \alpha_2$ (exactly as in *Step 2* above and Exercise 9) after which it is natural to interpret the endless

continued fraction we started with as representing the real number
$\alpha = \alpha_1 = \alpha_2$:

$$\alpha = [a_0; a_1, a_2, a_3, a_4, a_5, \ldots].$$

Let us now return to the example which first introduced us to the idea of
an *endless continued fraction*—namely the *attempt* to express the ratio
AD/BC (of the diagonal AD and the side BC of the regular pentagon
ABCDE) in the form of a continued fraction (Figure 37). Recall that

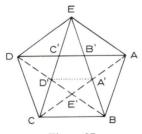

Figure 37

BC = B′D = AC′ and AB′ = A′D′, so that we have

$$\frac{AB}{BC} = \frac{B′D + AB′}{BC}$$

$$= 1 + \frac{1}{BC/AB′} = 1 + \frac{1}{AC′/AB′} = 1 + \frac{1}{(AB′ + B′C′)/AB′}$$

$$= 1 + \frac{1}{1 + \dfrac{1}{AB′/B′C′}} = 1 + \frac{1}{1 + \dfrac{1}{A′D′/B′C′}}$$

Since A′D′ and B′C′ are just the diagonal and side of the smaller pentagon,
we could also obtain a similar expression for A′D′/B′C′, and so the whole
process recurs. For this reason we were tempted to write

$$\frac{AD}{BC} = [1; 1, 1, 1, 1, 1, \ldots].$$

But this was at the time no more than a *flight of fancy*, and we should now
check to see whether it is compatible with the numerical meaning we have
just attributed to endless continued fractions. In other words we must show
that the real number, which is pinpointed by the endless sequence

$$[1], \quad [1; 1], \quad [1; 1, 1], \quad [1; 1, 1, 1], \quad [1; 1, 1, 1, 1], \ldots \quad (2.13.4)$$

of ordinary continued fractions, is precisely AD/BC.

Since AD/BC = A′D′/B′C′, it will be convenient to denote this ratio by
the Greek letter $\tau$ (pronounced "tor", or "tow" as in "towel"). Though it is
not strictly necessary, we shall make use of the slightly surprising, but very

elementary, geometrical fact (see Exercise 10) that

$$\frac{BC}{AB'} = \frac{AD}{BC} \qquad (= \tau).$$

If we now make use of this notation, then we can summarise our attempt to express AD/BC in the form of a continued fraction in a particularly simple way: namely

$$\frac{AD}{BC} = 1 + \frac{1}{\tau} = 1 + \cfrac{1}{1 + \cfrac{1}{\tau}} = 1 + \cfrac{1}{1 + \cfrac{1}{1 + \cfrac{1}{\tau}}} = \ldots .$$

It is now easy to show (see Exercise 11) that

$$[1] < [1; 1, 1] < [1; 1, 1, 1, 1] < \ldots < \frac{AD}{BC} < \ldots < [1; 1, 1, 1, 1, 1] < [1; 1, 1, 1] < [1; 1].$$

from which it is clear that the real number pinpointed by the sequence (2.13.4) is precisely AD/BC, and we can therefore happily write

$$\tau = \frac{AD}{BC} = [1; 1, 1, 1, 1, 1, \ldots].$$

EXERCISES

1. Given the set-up on page 138 with $C'D < A'B$ and

$$\frac{AB}{CD} = a_0 + \cfrac{1}{a_1 + \cfrac{C'D}{A'B}}$$

   show that

$$a_0 < a_0 + \frac{1}{a_1 + 1} < \frac{AB}{CD} < a_0 + \frac{1}{a_1} \leq a_0 + 1.$$

2. Given the set-up on page 139 with $A''B < C'D$ and

$$\frac{AB}{CD} = a_0 + \cfrac{1}{a_1 + \cfrac{1}{a_2 + A''B/C'D}}$$

   show that

$$a_0 < a_0 + \frac{1}{a_1 + 1} \leq a_0 + \cfrac{1}{a_1 + \cfrac{1}{a_2}} < \frac{AB}{CD} < a_0 + \cfrac{1}{a_1 + \cfrac{1}{a_2 + 1}} < a_0 + \frac{1}{a_1} \leq a_0 + 1.$$

3. (i) Express each of the following continued fractions as ordinary fractions (i.e. in the form $a/b$):

   (a) $1 + \cfrac{1}{2}, \quad 1 + \cfrac{1}{1 + \cfrac{1}{2}}, \quad 1 + \cfrac{1}{1 + \cfrac{1}{1 + 1/2}}, \ldots$

(b)  $1 + \dfrac{1}{2}, \quad 1 + \dfrac{1}{2 + \dfrac{1}{3}}, \quad 1 + \dfrac{1}{2 + \dfrac{1}{3 + 1/4}}, \ldots$

(c)  $1 + \dfrac{1}{2}, \quad 1 + \dfrac{1}{2 + \dfrac{1}{2}}, \quad 1 + \dfrac{1}{2 + \dfrac{1}{2 + 1/2}}, \ldots$

(ii) Square each fraction you obtained in (i)(c). What do you notice? [Hint: If necessary, use a calculator to express each of your answers as a decimal.]

(iii) Express each of the following continued fractions *first* as an ordinary fraction, and *then* as a decimal:

$$[3;7], \quad [3;7,15], \quad [3;7,15,1], \quad [3;7,15,1,292].$$

If

$$\alpha = [3;7,15,1,292,\ldots]$$

show, as in Exercise 1, that

$$3 < \alpha < 3\frac{1}{7}$$

and, as in Exercise 2, that

$$3\frac{15}{106} < \alpha < 3\frac{16}{113}$$

Go one step further to show that

$$3\frac{4687}{33102} < \alpha < \ldots .$$

(iv) Express each of the following ordinary fractions as (abbreviated) continued fractions:

$$7/11, \quad 91/143, \quad 8/3, \quad 49/18, \quad 487/179, \quad 55/34, \quad 4181/2584, \quad 3.14159.$$

4. Suppose that

$$a/b = [a_0 ; a_1, a_2, \ldots, a_n] \qquad \text{and} \qquad ra/rb = [b_0 ; b_1, b_2, \ldots, b_m].$$

Show that $a_0 = b_0$, and that

$$\text{if} \quad \frac{a}{b} = a_0 + \frac{a'}{b} = a_0 + \frac{1}{b/a'}$$

$$\text{then} \quad \frac{ra}{rb} = a_0 + \frac{1}{rb/ra'}.$$

Hence prove (by induction on $n$) that the continued fractions for $a/b$ and $ra/rb$ are identical.

5. (i) In each of the following pairs of continued fractions, find the largest:

(a) $[2;3], \quad [2;4]$

(b) $[2;3,4], \quad [2;3,5]$

(c) $[2; 3, 4, 5]$,   $[2; 3, 4, 6]$
(d) $[2; 3, 4, 5, 6]$,   $[2; 3, 4, 5, 7]$.

(ii) Complete the following assertions:

(a) $[a_0 ; a_1] > [a_0 ; b_1]$   precisely when $a_1 \ldots b_1$.
(b) $[a_0 ; a_1] > [b_0 ; b_1]$   precisely when $(a_0 \ldots b_0)$ or $(a_0 \ldots b_0$ and $a_1 \ldots b_1)$.
(c) $[a_0 ; a_1, a_2] > [a_0 ; a_1, b_2]$   precisely when $a_2 \ldots b_2$.
(d) $[a_0 ; a_1, a_2] > [b_0 ; b_1, b_2]$   precisely when $\ldots$.
(e) $[a_0 ; a_1, a_2, a_3] > [a_0 ; a_1, a_2, b_3]$   precisely when $a_3 \ldots b_3$.
(f) $[a_0 ; a_1, a_2, a_3] > [b_0 ; b_1, b_2, b_3]$   precisely when $\ldots$.

(iii) Let $\alpha = [a_0 ; a_1, \ldots, a_m]$     and     $\beta = [b_0 ; b_1, \ldots, b_n]$.
We have seen in the text that

$$a = \beta \text{ precisely when } m = n \text{ and } a_i = b_i \text{ for each } i \ (0 \leq i \leq m).$$

*We want to know exactly when $\alpha < \beta$.* To allow for the possibility that $m \neq n$, we put $\infty = a_{m+1} = a_{m+2} = \ldots$ and $\infty = b_{n+1} = b_{n+2} = \ldots,$[4]

$$\alpha = [a_0 ; a_1, a_2, \ldots, a_m, \infty, \infty, \ldots] \quad \text{and} \quad \beta = [b_0 ; b_1, b_2, \ldots, b_n, \infty, \infty, \ldots].$$

Prove (by induction on the smaller of $m$ and $n$) that

$$\alpha < \beta \text{ precisely when} \begin{cases} a_0 = b_0, \ a_1 = b_1, \ldots, \ a_{j-1} = b_{j-1}, \text{ but } a_j \neq b_j \text{ for some } j, \\ \text{with } a_j < b_j \text{ if } j \text{ is even, and } a_j > b_j \text{ if } j \text{ is odd.} \end{cases}$$

6. Express the first ten terms of the endless sequence

$$[1], \quad [1; 2], \quad [1; 2, 3], \quad [1; 2, 3, 4], \quad [1; 2, 3, 4, 5], \quad [1; 2, 3, 4, 5, 6], \quad \ldots$$

*first* as ordinary fractions and *then* as decimals. Do they appear to be heading towards some real number $\alpha$?

7. Let

$$a_0, a_1, a_2, a_3, a_4, a_5, \ldots$$

be any real numbers with $a_0 \geq 0$ and all the other $a_i$'s $> 0$.

(i) Prove each of the following three statements (by induction on $n$):

(a) $[a_0 ; a_1, a_2, \ldots, a_{2n}] < [a_0 ; a_1, a_2, \ldots, a_{2n+2}]$   (for all $n \geq 0$);
(b) $[a_0 ; a_1, a_2, \ldots, a_{2n-1}] > [a_0 ; a_1, a_2, \ldots, a_{2n+1}]$   (for all $n \geq 1$);
(c) $[a_0 ; a_1, a_2, \ldots, a_{2n}] < [a_0 ; a_1, a_2, \ldots, a_{2m+1}]$   (for all $m, n \geq 0$).

(ii) Deduce that

$$[1] < [1; 2, 3] < [1; 2, 3, 4, 5] < \ldots < [1; 2, 3, 4, 5, 6] < [1; 2, 3, 4] < [1; 2].$$

8. The fundamental property of real numbers (Chapter II.10) applies to any *increasing* sequence

$$a_1, a_2, a_3, a_4, a_5, \ldots$$

whose terms are all *less than or equal to* some number $K$.

---

[4] We assume that $\infty$ satisfies $1/\infty = 0$, $\infty + 0 = \infty$, and $k < \infty$ for every whole number $k$.

Suppose we have a *decreasing* sequence

$$b_1, b_2, b_3, b_4, b_5, \ldots$$

whose terms are all *greater than or equal to* some number $L$.

(i) Use this to construct a closely related *increasing* sequence, to which we can apply the fundamental property of real numbers. [Hint: A sequence of *negative* terms can still be increasing!]

(ii) Deduce that a *decreasing* sequence, whose terms are all *greater than or equal to* some number $L$, is necessarily heading towards a real number $\beta \geq L$.

9. (i) (a) Let $K_1 = [a_0 ; a_1]$ and $K_2 = [a_0]$. Show that $K_1 - K_2 = 1/a_1$.
   (b) Let $L_1 = [a_1 ; a_2]$ and $L_2 = [a_1 ; a_2, a_3]$. Show that $L_1 - L_2 < 1/a_2^2 a_3$.
   (c) Let $K_1 = [a_0 ; a_1, a_2, a_3]$ and $K_2 = [a_0 ; a_1, a_2]$. Show that

$$K_1 - K_2 < 1/a_1^2 a_2^2 a_3.$$

   (d) Let $L_1 = [a_1 ; a_2, a_3, a_4]$ and $L_2 = [a_1 ; a_2, a_3, a_4, a_5]$. Show that

$$L_1 - L_2 < 1/a_2^2 a_3^2 a_4^2 a_5.$$

(ii) Assume that *both* the following inequalities are true:

$$[a_0 ; a_1, \ldots, a_{2n-1}] - [a_0 ; a_1, \ldots, a_{2n-2}] < \frac{1}{a_1^2 a_2^2 \ldots a_{2n-2}^2 a_{2n-1}}$$

$$[a_1 ; a_2, \ldots, a_{2n-2}] - [a_1 ; a_2, \ldots, a_{2n-1}] < \frac{1}{a_2^2 a_3^2 \ldots a_{2n-2}^2 a_{2n-1}}.$$

Use these two statements to prove that *both* the following are then also true:

$$[a_0 ; a_1, \ldots, a_{2n+1}] - [a_0 ; a_1, \ldots, a_{2n}] < \frac{1}{a_1^2 a_2^2 \ldots a_{2n}^2 a_{2n+1}}$$

$$[a_1 ; a_2, \ldots, a_{2n}] - [a_1 ; a_2, \ldots, a_{2n+1}] < \frac{1}{a_2^2 a_3^2 \ldots a_{2n}^2 a_{2n+1}}.$$

(iii) Deduce that the inequality

$$[a_0 ; a_1, \ldots, a_{2n-1}] - [a_0 ; a_1, \ldots, a_{2n-2}] < \frac{1}{a_1^2 a_2^2 \ldots a_{2n-2}^2 a_{2n-1}}$$

holds for every $n \geq 1$.

(iv) Let $K_1 = [1 ; 2, 3, \ldots, 2n]$ and $K_2 = [1 ; 2, 3, \ldots, 2n - 1]$.
Use (iii) to show that

$$K_1 - K_2 < \frac{1}{2^2 \cdot 3^2 \cdot \ldots \cdot (2n - 1)^2 \cdot 2n}.$$

Hence show that $\alpha_1 = \alpha_2$ as claimed on page 147.

10. Show that, in Figure 37

$$\frac{BC}{A'D'} = \frac{BE}{A'E}.$$

Hence deduce that

$$\frac{BC}{AB'} = \frac{AD}{BC}.$$

11.  (i) Apply the result of Exercise 7(i)(a) to show that

$$[1] < [1; 1, 1] < [1; 1, 1, 1, 1] < [1; 1, 1, 1, 1, 1, 1] < \ldots .$$

(ii) Apply the result of Exercise 7(i)(b) to show that

$$\ldots < [1; 1, 1, 1, 1, 1] < [1; 1, 1, 1] < [1; 1].$$

(iii) Use the result of Exercise 7(i)(c), and the fact that we may write

$$AD/BC = [1: \tau] = [1; 1, \tau],$$

to show that

$$[1] < [1; 1, 1] < [1; 1, 1, 1, 1] < \ldots < AD/BC <$$
$$\ldots < [1; 1, 1, 1, 1, 1] < [1; 1, 1, 1] < [1; 1].$$

12. Show that $\tau = \dfrac{1 + \sqrt{5}}{2}$.

13. In Chapter II.4 we discussed in detail the attempt to find the greatest common measure of the diagonal AC and the side BC of the square ABCD (Figure 38).

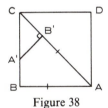

Figure 38

Working as on page 148, express the ratio AC/BC as an endless continued fraction. [Hint: The procedure begins thus:

$$\frac{AC}{BC} = \frac{AB' + B'C}{BC} = 1 + \frac{1}{BC/B'C} = \ldots]$$

14. Let

$$\alpha = [1; 1, 2, 1, 1, 4, 1, 1, 6, 1, 1, 8, 1, 1, 10, 1, 1, \ldots].$$

Evaluate (as decimals) the first few terms of the corresponding endless sequence

$$[1], [1; 1], [1; 1, 2], [1; 1, 2, 1], [1; 1, 2, 1, 1], [1; 1, 2, 1, 1, 4], \ldots$$

then try to identify the real number $\alpha$. [Hint: Consider $1 + \alpha$.]

15. Find some way of working out the endless continued fractions corresponding to

$$\sqrt{3}, \quad \sqrt{5}, \quad \sqrt{6}, \quad \sqrt{7}, \quad \sqrt{8}.$$

16. If your calculator has square root ($\sqrt{x}$) and reciprocal ($1/x$) keys, try the following.

Enter a fraction $p/q$ (for the first few trials try (a) $p = 3$, $q = 1$, (b) $p = 5$, $q = 1$, (c) $p = 6$, $q = 1$, (d) $p = 7$, $q = 1$; then try other values of $p$ and $q$).

(0) Press $\sqrt{x}$.
(1) Write down the integer part of the result.
(2) Subtract it off.
(3) Press $1/x$.
(4) Write down the integer part of the result.
(5) Subtract it off.
(6) Press $1/x$.
(7) Write down the integer part of the result.
(8) Subtract it off.
(9) Press $1/x$.
   Etc.

What do you notice about the sequence of integers you write down each time? What has this to do with continued fractions?

17.  (i) Show that

$$[1; 1, 1, 1, \ldots]^2 = [2; 1, 1, 1, \ldots]$$
$$[1; 2, 2, 2, \ldots]^2 = [2].$$

(ii) Show that

$$[1; 1, 1, 1, \ldots] \times [0; 1, 1, 1, \ldots] = [1]$$
$$[2; 2, 2, 2, \ldots] \times [0; 2, 2, 2, \ldots] = [1].$$

Generalise (and prove your generalisation).

(iii) Check and complete the following.

$$[1; 2] + [1; 2] = [2] \times [1; 2] = [3].$$
$$[1; 2, 3] + [1; 2, 3] = [2] \times [1; 2, 3] = [2; 1, 6].$$
$$[1; 2, 3, 4] + [1; 2, 3, 4] = [2] \times [1; 2, 3, 4] = \underline{\qquad}.$$
$$[1; 2, 3, 4, 5] + [1; 2, 3, 4, 5] = [2] \times [1; 2, 3, 4, 5] = \underline{\qquad}.$$

# PART III
# GEOMETRY

In which
- we analyse the notions of 2-dimensional area, 3-dimensional volume, the length of a curve, and the area of a curved surface, together with the infinite processes on which they depend;
- we discover that any attempt to invest these elementary geometrical ideas with precision involves the fundamental property of real numbers in an essential way, and that, though the intuitive ideas of area and tangents may aid our naive understanding of the calculus, in their precise form these elementary geometrical ideas become exceedingly complicated—and even seem to depend on the calculus itself;
- and we begin to see why the 1870 version of the calculus insists on the logical separation of the calculus from geometry.

# Numbers and Geometry

In Parts I and II we have gone out of our way to stress the enormous difference between the *finite* procedures of ordinary arithmetic, and those mathematical concepts whose very meaning depends on *the introduction and interpretation of infinite processes.* In contrast, you have in the past been encouraged to use real numbers (whether rational or irrational) in a naive, unquestioning way—especially in geometry: for example, you have been quietly encouraged to assume that, if we measure the length of a line segment AB in terms of some given *unit segment* CD, then its length AB/CD can obviously be expressed as a real number. While this *is* obvious when CD fits into AB a whole number of times leaving no remainder, or when CD and AB have some *common measure* MN which fits into CD precisely $b$ times with no remainder and into AB precisely $a$ times with no remainder (in which case AB/CD $= a/b$), it is not at all obvious in general. In Chapter II.13 we saw one way of justifying the belief that AB can always be measured in terms of CD, but it was not exactly obvious!

In Part III we shall look at other ways in which numbers arise in geometry (2-dimensional area, 3-dimensional volume, length of curves in 2- and 3-dimensions, curved surface area in 3-dimensions) with the slightly perverse intention of convincing you that the relationship between numbers and geometry is much more awkward than you had previously realised. You will then hopefully begin to appreciate just why the 1870 cleaned-up version of the calculus *deliberately avoids* all reference to geometry— preferring first to sort out precisely what we mean by real numbers, and then to develop the calculus purely in terms of these real numbers.

We have already mentioned the way in which the arithmetic of the ancient Greeks became logically and psychologically embedded in geometry. Thus the Pythagoreans had separate notions of

*number* (that is, *whole numbers and their ratios*), and *geometry*.

They "sensed"

that relationships in geometry could be expressed by numbers, and
that relationships between numbers should have consequences
in geometry.

But their notion of number was restricted to *whole numbers and their ratios*, and could therefore only express relationships *between pairs of line segments* (*or areas*) *which have a common measure.* They then discovered that some naturally occurring pairs of line segments, such as the side and diagonal of a square or a regular pentagon, have no common measure. This put paid to the idea that mathematics could be constructed on the twin supports of

*number*    and    *geometry.*

Consequently these two basic ingredients of mathematics were forcefully separated *for the first time.* Admittedly it was *sometimes* possible to express geometrical relationships in terms of (whole) numbers, but there were many cases in which no such numerical description seemed to be possible. So since the Pythagorean notion of number was only capable of expressing *some* of the relationships to be found in geometry, *geometry* was felt to be mathematically more far-reaching than *number.* As a result, later Greek mathematics is essentially geometrical: it worked with *magnitudes* and *squares,* but its *magnitudes* were genuine geometrical line segments and rectangles rather than lengths and areas; its *squares* were genuine geometrical squares rather than products of one number by itself. Thus, for Euclid (around 300 BC), Pythagoras' theorem was *not* a numerical statement about squares of lengths, but a statement about genuine geometrical squares: *the square* [drawn] *on the hypotenuse* [can be cut up and rearranged so that it] *is equal to the sum of the two squares* [drawn] *on the other two sides.*

In spite of this apparent handicap, Greek mathematicians developed a remarkably clever theory of purely geometrical relationships which is in some sense a geometrical version of our real numbers.[1] You may get some feeling for the conceptual complexity of their approach from the fact that where we refer blithely to

"$\sqrt{2}$"

they had to find some way of referring to

> "*all pairs of segments (AB, CD) which bear the same relationship to one another as do the diagonal and side of a square.*"

Moreover, they had to find some way of adding, multiplying, and comparing these geometrical "number-substitutes."

The more one understands the difficulties the Greeks had to overcome, the more impressive their achievements seem. But by the middle of the seventeenth century the reasons behind their conscious separation of *number* and *geometry* were either no longer appreciated, or no longer felt to

---

[1] The real numbers can be constructed from the rational numbers by means of *Dedekind cuts*: in contrast to infinite decimals, this approach makes no overt use of infinite processes. Dedekind got the idea for his "cuts" from the Greeks, who were exceedingly reluctant to use infinite processes (at least in public)—apparently because they appreciated the difficulty of giving a strictly logical justification for their use.

be as convincing as they had once seemed. The distinction between *number* and *geometry* had in no sense been resolved; but it had become blurred partly as a result of mathematicians' increased confidence in dealing with numbers, and partly as a result of their growing preference for practical exploration and experiment as opposed to rigorous philosophising. And although mathematicians in the sixteenth and seventeenth centuries were greatly inspired by the rediscovery of Greek mathematical texts, they did not share the Greeks' philosophical qualms about the fact that the presumed connection between numbers and geometry could not be strictly justified.[2]

By the mid seventeenth century numerical calculation had become relatively straightforward—at least in an algorithmic sense: the decimal notation was now standard, even for decimal fractions; numerical calculation had become much more important in commerce, navigation, architecture, etc; Napier's logarithms appeared in 1614 and were soon simplified and widely used; the square root sign was used unquestioningly for rationals and irrationals; and algebra flourished. Descartes (1596–1650) was understandably delighted when he realised (around 1630) that, by introducing axes and coordinates, he could use elementary arithmetic and algebra to "prove" many propositions which the ancient Greeks had found either very hard or impossible to prove. But Descartes never bothered to explain *why* *numerical* coordinates, *arithmetic*, and *algebra* could be used to prove *geometrical* theorems.

It would be misguided to criticise Descartes, for it was from the subsequent fruitful confusion that the calculus emerged (around 1670). Neither should we be surprised to find mathematical ideas, like coordinate geometry, beginning life in a half-baked state—though ideally they should not stay that way for very long. In this instance the precise nature of real numbers was not clarified until 1859–72 by Dedekind, Weierstrass (1815–1897) and Cantor (1845–1918), and an explicit justification of the use of real numbers in geometry had to wait until Hilbert (1862–1943) delivered his famous series of lectures on the *Foundations of Geometry* in 1898/9. A careful analysis of the way real numbers are used in geometry necessarily raises awkward questions about the nature of real numbers, about the nature of geometry, and about the relationship between real numbers and geometry. These awkward questions were eventually resolved in the nineteenth century

(i) by constructing the real numbers from the whole numbers in a purely *arithmetical* way, and

---

[2] It is not even clear to me that mathematicians in the seventeenth century were aware of the problem. To us, the fact "that each point of a line ... corresponds to a real number and conversely ... represents a remarkable link between something which is given by our spatial intuition and something that is constructed in a purely logico-conceptual way." (H. Weyl, *The Continuum*, Berlin, (1918)).

(ii) by basing the notions of function, limits, and the whole of the calculus
    on these *arithmetical* real numbers—rejecting all use of geometrical ar-
    guments.

Thus number and geometry were forcefully separated *for the second time.*

The purely *logical ingenuity* required to separate calculus from geometry,
and the *psychological effort* involved in doing without geometrical arguments make considerable demands on all students. A few may be content to
experience the *pattern-maker's delight* at being able to build up so much
mathematics from so simple a starting point. But most, having learned to
use *carefree calculus* with no apparent bother, will need to keep asking
"Why? Why? Why?" if they are eventually to understand the reason for
starting to build the calculus all over again from scratch.

As we saw in Part I, starting again from scratch is a common feature of
all learning. A consistent arithmetical approach to real numbers *is certainly
possible*, and 2- and 3-dimensional euclidean geometry *can be reduced* to the
study of ordered pairs and ordered triples of real numbers; but, no matter
how inefficient it may seem, one simply cannot explain the details to a
beginner, who has no choice but to begin by mastering those experiences
relevant to the beginner—though she/he may later wish to reconsider their
precise logical justification. For the beginner, numbers inevitably start life
as *numbers of "things"* (say *sweets*)—with addition and subtraction arising
in the form of *combining* and *taking away*; multiplication and division may
arise by considering *m lots of n sweets*, or *sharing n sweets between m
people*—in both of which one experiences an asymmetry between the two
numbers *m* (*lots* or *people*) and *n* (*sweets*) which will eventually have to be
unlearnt. Fractions may arise by extending these ideas of sharing to *sharing
one cake between four people*: and though *parts of a cake* lend themselves
fairly naturally to addition and subtraction, the multiplication and division
of fractions presents us with a fresh challenge. This challenge is met rather
lamely by exploiting that little word *of*: since $m \times n$ first arose by consider-
ing *m lots of n things*, so $\frac{3}{4} \times \frac{2}{3}$ is declared to mean $\frac{3}{4}$ *of* $\frac{2}{3}$. Division of
fractions fares even worse! Negative numbers appear in yet another
disguise—as *debts* or *steps backwards*; but however they are introduced,
addition can be handled more or less as before, though subtraction
$(+3 - (-4) = ?)$ is a little harder. However, when we come to multiply and
divide negative numbers, it is no longer possible to escape from the fact that
*every time we extend the scope of the word number, we should* (strictly speaking) *revise the way we think of "numbers"*: first *sweets*, then *lots* and *people*,
then *bits of cake* and *bits of bits* (and *bits divided by bits!*), then *steps
backward* and *debts*. They were all so harmlessly called *"numbers"*, and the
operations kept the same familiar names $+, -, \times, \div$: but we have in fact
by this stage had to master many different number concepts.

This variety is something of an embarrassment to the mathematician,
who would much prefer a single abstract mental picture of number. The

first really effective synthesis of this kind comes when we begin to picture numbers on *the number line*. In retrospect we may describe the process in the following way: first one chooses an arbitrary *point* on the line—call it A (Figure 39),

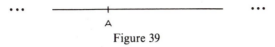

A

Figure 39

then *a positive direction* (usually to the right), and *a basic unit segment* AB in that direction (Figure 40).

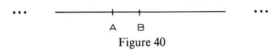

A    B

Figure 40

One can then call the points A and B simply 0 and 1 (Figure 41).

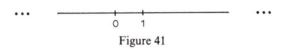

0    1

Figure 41

By shifting this basic unit segment *to the right* we can represent all the (positive) whole numbers as points on the line (Figure 42).

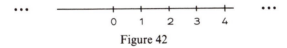

0   1   2   3   4

Figure 42

Next the positive fractions are just asking to be filled in; and the yawning gap on the left simply begs us to shift our unit segment not just to the right, but also *to the left* (Figure 43).

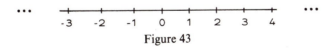

-3   -2   -1   0   1   2   3   4

Figure 43

When $\sqrt{2}$ appears, it too can be slotted in. And when the time comes to imagine complex numbers, we need only step from the 1-dimensional *number line* to the 2-dimensional *complex plane*.

The organising power of this picture of real numbers is very impressive: it is one of those true master-strokes which make advanced mathematics suddenly appear elementary—like our Hindu-Arabic numeral system, or the use of letters in algebra. But for our purposes it is simply not good enough.

# The Role of Geometrical Intuition

> ... the sketch which suffices for physics is not the deduction
> which analysis requires. But it does not follow thence that
> one cannot aid in finding the other.
> Henri Poincaré, *The Value of Science*

The fact that our mental picture of *real numbers* is all tied up with the *geometrical* number line certainly helps our *intuitive* understanding of their properties: inequalities ($a < b$) correspond naturally to the relative position of points on the number line (to the left of/to the right of); addition and subtraction correspond to *shifts* (to the right and to the left); and multiplication ($\times a$) can be thought of either in terms of *enlarging* (*a* times), or in terms of areas of rectangles. It makes good psychological sense to choose a framework for real numbers which exploits our *intuitive geometrical understanding* of points on a line—much of which derives from our experience of drawing and measuring with a ruler. But when the time comes to examine the precise nature of real numbers, this kind of dependence on geometrical intuition is logically indefensible, *unless geometry itself can be shown to be in some sense a simpler, or a more natural starting point*—in which case we should instead begin by analysing the details of our geometrical intuition as carefully as we can. The examples we discuss in Part III are intended to convince you that in a first course in analysis it is simpler to abstain completely from the use of geometrical arguments, and to develop both the real numbers and the calculus itself *purely arithmetically*.

The influence of geometrical intuiton is by no means restricted to our mental picture of real numbers. Most of us were introduced to the idea of *the derivative of $f(t)$* (at $t = t_0$) in terms of *the slope of the graph of $f(t)$*(at $t = t_0$), or in terms of *the instantaneous velocity* (at time $t_0$) of a particle whose position on the *y*-axis at time *t* is given by the single coordinate $f(t)$. *The integral of $f(t)$* (between $t = a$ and $t = b$) is also generally thought of as *the area under the graph of $f(t)$*(between $t = a$ and $t = b$). Thus the two fundamental ideas of the calculus are usually introduced in a way which makes them look as though they were *essentially geometrical* (or *dynamical*)—and they are introduced in this way for two very good reasons:

(1) Firstly, many of the most impressive and most useful elementary applications of the calculus occur when it is used to solve problems in

geometry and in dynamics, and these applications are much easier to understand if the fundamental ideas of the calculus are themselves introduced in a geometrical or a dynamical spirit.[1]

(2) Secondly, just as any awkwardness we might feel about negative and irrational numbers, or about complex numbers, is greatly reduced if we allow ourselves to think of them geometrically as points on the (real) number line or in the (complex) plane, so the fundamental ideas of the calculus appear to be easier to understand if we appeal to geometrical pictures of tangents to curves, and areas under curves, and to intuitive physical notions of velocity and acceleration.

The first of these two reasons is purely *tactical*: we are introduced to the calculus in this way in the hope that we will find the important elementary applications easier to swallow. The second reason is purely *psychological*: when introducing new and abstract ideas it is helpful to present them in terms of relatively familiar, actually visible, or intuitively obvious examples. *Neither reason has anything strictly to do with mathematics.* When we are *groping* towards our first notions of derivative and integral these geometrical and dynamical ideas are useful precisely because of the richness of our *experience and intuition* of space and motion. *But this experience and intuition*, which helps us psychologically to accept the basic ideas of the calculus, *is a positive hindrance when it comes to understanding its logical, or mathematical, justification.* It would be wrong to conclude that logic is the pedant's substitute for experience and intuition. **We all need experience and intuition to generate new ideas, but we have to learn that ideas based on intuition need to be very carefully analysed, and cannot simply be taken on trust.**

In the remaining chapters of Part III we shall see that *some* of the *intuitive* elementary geometrical ideas which you might be inclined to call obvious, are in fact much more difficult to justify than you might have suspected; and that others are simply false. The conclusion we shall draw is simply this: *However valuable geometrical intuition may be in formulating our first ideas about the calculus, when it comes to understanding the strict mathematical justification of the calculus, geometrical ideas are more trouble than they are worth.*

---

[1] Since the most important use of real numbers in school mathematics occurs in connections with graphs, coordinate geometry and the solution of equations, this same observation applies equally as a reason for encouraging students *initially* to think of real numbers in terms of the real number line.

CHAPTER III.3

# Comparing Areas

In Part II we considered the problem of *measuring the length of line segments*—or, to be more precise, we considered the problem of *comparing* two line segments AB, CD *by looking for a common measure.* We discovered that a given pair of line segments AB, CD may have no common measure at all. However in this case, though our procedure for finding common measures could not possibly produce a genuine common measure as it was designed to do, it could nevertheless be interpreted to obtain a value for, or a sequence of better and better approximations to, the *irrational* ratio AB/CD (see Chapter II.13).

In Chapter II.1, when discussing Pythagoras' theorem, we also *measured areas of squares (and triangles)*—or, to be more precise, we *compared* squares drawn on a square dot-lattice by measuring each one in terms of one of the basic unit squares—as for example in Figure 3 of Chapter II.1 (Figure 44). Fortunately for us, the squares we compared in Chapter II.1 always worked out *exactly* as a *whole number* of unit squares: in other words, *we only tried to compare squares which had the basic unit square as a common measure.* But most of the time the shapes which we wish to compare in 2-dimensions will have no such common measure; moreover, this lack of a common measure can arise in two rather different ways.

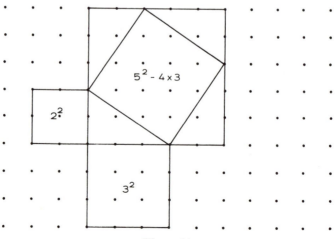

Figure 44

164

Firstly, two shapes A and B can certainly have no common measure *if the ratio*

$$\frac{area\ A}{area\ B}$$

*is irrational.* This is because if shapes A and B have a common measure C (Figure 45), then C must fit exactly a whole number of times (say *a* times)

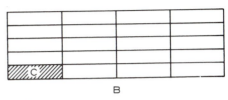

A                                      B

Figure 45

into shape A, and a whole number of times (say *b* times) into shape B, so that

$$\frac{area\ A}{area\ B} = \frac{a}{b}$$

and so this ratio must be a rational number. From this observation it follows, for example, that no two of the shapes in Figure 46 can have a common measure.[1]

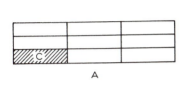

Figure 46

---

[1] Although we have proved that $\sqrt{2}$ is *irrational*, we have not yet proved the familiar formula

$$length \times breadth$$

for the area of a rectangle, so we should not really use this to calculate (area Y/area X). We can nevertheless show by another method that

$$\frac{area\ Y}{area\ X} = \frac{\sqrt{2}}{1}$$

(see Exercise 1). Hence shapes X and Y can have no common measure. In the case of shape Z we have not yet proved the familiar formula

$$\pi r^2$$

for the area of a circle of radius $r$ (see Exercise 2). Neither have we proved that $\pi$ and $\pi/\sqrt{2}$ are *irrational* (Exercise 3). Thus, strictly speaking, we are not yet really in a position to show that

$$\frac{area\ Z}{area\ X} = \pi \quad and \quad \frac{area\ Z}{area\ Y} = \frac{\pi}{\sqrt{2}}.$$

But once these have been proved, we could then use the result of Exercise 2 to conclude that shapes Z and X (resp. shapes Z and Y) have no common measure.

Secondly, *even if the ratio (area A/area B) happens to be a rational number, the two shapes A and B will usually have no common measure.* For example, if A is a "1 by 1" square and B is an "$\alpha$ by $1/\alpha$" rectangle, then area A = area $B^2$ so (area A/area B) = 1 is certainly rational, but one would not in general expect to find a single shape C which fits a whole number of times into both A and B; (however, this may be possible for certain special values of $\alpha$—thus it is not hard to find a triangle C which fits exactly eight times into each of the shapes A and B in Figure 47).

Figure 47

There are other complications in 2-dimensions which simply do not arise in 1-dimension. For example, whereas for two line segments AB, CD *which have no common measure* our procedure of repeatedly subtracting the smaller segment from the larger provided a way of calculating the (irrational) ratio AB/CD, we cannot hope to find an analogous procedure which will work in 2-dimensions for the following simple reason. Let AB, CD be two line segments with CD < AB, and let X, Y be two shapes in 2-dimensions. The two line segments AB, CD are necessarily *similar in shape :* and given two similar shapes, one can always subtract the smaller shape from the larger leaving a remainder (Figures 48 and 49). Now for the two line seg-

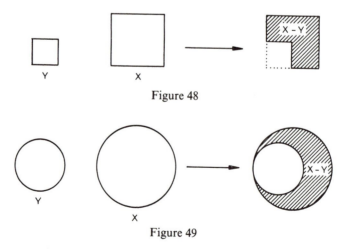

Figure 48

Figure 49

---

[2] As stated in the previous footnote, we are not yet in a position where we can simply use the *length × breadth* formula to show that *area A = area B.* Two alternative approaches are described later on in this section.

ments AB and CD the remainder AB − CD just happens to be similar in shape to CD, so that we can repeat the whole process with the two segments AB − CD and CD in place of AB and CD. And so it goes on. In 2-dimensions the two shapes X and Y we start with will usually not be similar to one another. If they are *not* similar in shape, then we must be prepared to find that neither shape fits inside the other, so that it is simply not possible to subtract one from the other. And even if the initial shapes X and Y *are* similar in shape (with X larger than Y) then it seems most unlikely that X − Y will ever be similar in shape to Y. Hence the procedure simply peters out.

It now looks as if working with 2-dimensional shapes is going to be more complicated than working with line segments! In the rest of this chapter we shall explore two different approaches to the problem of comparing areas in 2-dimensions.

The first approach is an attempt to imitate the principle of the ruler. By applying a ruler to the segment AB we may tell by eye, that its length is, for example,

greater than 6 cms (but less than 7 cms),
greater than 6.2 cms (but less than 6.3 cms);

and if we have some means of magnifying the scale, then we may continue in this way to pinpoint the true length of the segment AB more and more accurately. The 2-dimensional analogue of this procedure requires one to decide how many whole square centimetres (square millimetres, etc.) one can *fit completely inside* the given shape, and how many whole square centimetres (square millimetres, etc.) are needed to *totally cover* the given shape (Figure 50). We shall refer to this as the method of *inner and outer*

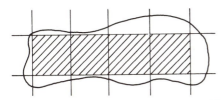 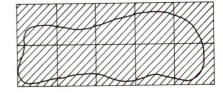

Figure 50

*approximations*. It is, in practice, much more awkward than measuring in 1-dimension, because the error in each measurement is spread all round the edge of the original shape, whereas in measuring the segment AB we were only worried about the error at the single point B. As it stands, the method also has theoretical limitations.[3] It can nevertheless be used to measure a reasonably large class of shapes.

---

[3] However we shall mostly hint at these without going into details—but see Exercise 7(iii).

Our second approach is based on the simple idea of *dissection*: if shape A can be dissected in such a way that the pieces can be reassembled to make shape B, then the two shapes A and B have equal areas. A very famous application of this idea occurs in the dissection proof of Pythagoras' theorem given by the Hindu mathematician Bhāskara (1114–1185?) but discovered much earlier in China. Start with any right-angled triangle with sides $a$, $b$, $c$ (where $a \leq b \leq c$). One can then fit four copies of this triangle together to make a $c$ by $c$ square with a $(b - a)$ by $(b - a)$ square hole in the middle (Figure 51). The area of the big square $(=c^2)$ must therefore be

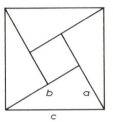

Figure 51

equal to the area of the square hole $\left(=(b - a)^2\right)$ plus the area of the four triangles $(= 2ba)$.

$$c^2 = (b - a)^2 + 2ba$$

$$\therefore c^2 = b^2 + a^2.$$

In the remainder of the chapter we shall consider how each of these approaches can be used to answer the following basic questions: *How can one compare the areas*

(1) *of two arbitrary rectangles ABCD and A'B'C'D'?*
(2) *of two arbitrary polygons?*
(3) *of two arbitrary shapes in the plane?*

The approach in terms of common measures and approximations is in some ways less attractive than the purely geometrical approach through dissection—for it uses the full force of the *fundamental property of real numbers* (Chapter II.10) from the very start, even when answering the simplest of our three questions ("How can one compare two rectangles?"). But this approach has one genuine virtue: namely, that it suggests strongly, from the very beginning, that *any geometrical theory of 2-dimensional area is, logically speaking, an application of the arithmetical properties of endless sequences of ordinary real numbers.* If this is in fact the case, then in our attempt to establish the calculus on a satisfactory basis, we should not be misled into thinking that by appealing to geometrical concepts like *area*, we can somehow avoid the arithmetical complexity of the fundamental property of real numbers. Indeed it may be both simpler and safer to replace each reference to *area* by some purely arithmetical statement about real numbers.[4]

But let us now use the method of inner and outer approximations to tackle the first of our three questions. We start with the naive idea on which all practical measurement is based: namely that *to compare rectangles ABCD and A'B'C'D', we need only measure each one in terms of some common unit rectangle WXYZ, and then compare the answers.* Now

> *if* the sides AB, A'B' are simple multiples of WX,
> and the sides AD, A'D' are simple multiples of WZ (Figure 52),
> *then* this procedure is both eminently reasonable and very effective

because

$$\text{ABCD contains precisely } \frac{AB}{WX} \cdot \frac{AD}{WZ} \text{ copies of the rectangle WXYZ}$$

and

$$\text{A'B'C'D' contains precisely } \frac{A'B'}{WX} \cdot \frac{A'D'}{WZ} \text{ copies of the same rectangle WXYZ}$$

so

$$\frac{\text{area ABCD}}{\text{area A'B'C'D'}} = \frac{AB}{A'B'} \cdot \frac{AD}{A'D'}.$$

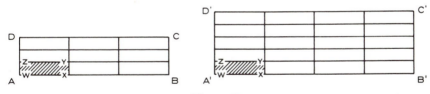

Figure 52

We can use essentially the same approach if we drop the idea of a *fixed* common measure WXYZ, and try instead to choose WX and WZ to suit

---

[4] In a similar way, any serious attempts to analyse the geometrical notions of

> *the continuity of a curve,*
> or       *the tangent to a curve at a given point*

or even the notion of

> *the curve itself*

are likely to lead to the conclusion that, for the purposes of establishing the calculus on a satisfactory basis, it might be wiser to replace each reference to such ideas by some purely arithmetical statement (see Exercise 4).

the particular pair of rectangles ABCD and A'B'C'D' which are being compared: thus

*if* AB and A'B' have a common measure (say WX),
and AD and A'D' have a common measure (say WZ) (Figure 53),
*then* we can construct a rectangle WXYZ which is a common measure for ABCD and A'B'C'D'.

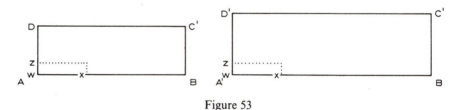

Figure 53

The rectangle WXYZ fits into ABCD precisely (AB/WX) · (AD/WZ) times, and into A'B'C'D' precisely (A'B'/WX) · (A'D'/WZ) times, so as before we have

$$\frac{\text{area ABCD}}{\text{area A'B'C'D'}} = \frac{AB}{A'B'} \cdot \frac{AD}{A'D'}.$$

In particular, if A'B'C'D' is a *unit square*, and AB, A'B' have a common measure WX, and if AD, A'D' have a common measure WZ—if, that is, the lengths AB and AD are *rational*—then

$$\text{area A'B'C'D'} = 1 \quad \text{and} \quad A'B' = A'D' = 1$$

and

$$\text{area ABCD} = AB \times AD$$

so we have proved the familiar *length × breadth* formula for *rectangles with rational sides*.

But if you have worked through any of the chapters in Part II, then you will realise that we cannot in general expect AB and A'B' (or AD and A'D') to have a common measure. Unfortunately it is not nearly so easy to prove either

$$\frac{\text{area ABCD}}{\text{area A'B'C'D'}} = \frac{AB}{A'B'} \cdot \frac{AD}{A'D'}$$

or the *length × breadth* formula

$$\text{area ABCD} = AB \times AD$$

when one factor (or both) on the right hand side is *irrational*.

Suppose, for example, that we wish to compare the two rectangles ABCD ("$\sqrt{2}$ by $\sqrt{3}$") and A'B'C'D' ("16/13 by 2") (Figure 54). The seg-

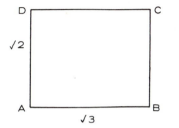 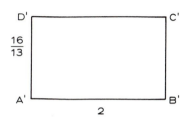

Figure 54

ments AB and A'B' here have no common measure; and the same is true of AD and A'D' (Exercise 5). So how exactly should we proceed?

Our present position may remind you of the difficulty we faced in Chapters II.5, II.6 and II.7. Then we were not comparing rectangles but were simply trying to express each and every real number $\alpha$ as a decimal—that is, in terms of positive and negative powers of 10. We found this to be entirely straightforward as long as the number $\alpha$ was itself a decimal fraction; but to handle numbers which were not of this form, such as 1/13 or $\sqrt{2}$, we had to introduce the idea of an *endless sequence of ordinary decimal approximations*—first in units, then in tenths, then in hundredths, and so on. You may recall that in the case of 1/13 we obtained the inequalities

$$0 < \frac{1}{13} < 1$$

$$0.0 < \frac{1}{13} < 0.1$$

$$0.07 < \frac{1}{13} < 0.08$$

$$0.076 < \frac{1}{13} < 0.077$$

$$0.0769 < \frac{1}{13} < 0.0770$$

$$\vdots$$

We can now use the same idea to obtain inequalities for the areas of our "$\sqrt{2}$ by $\sqrt{3}$" rectangle ABCD and our "16/13 by 2" rectangle A'B'C'D'.

We start by measuring each one using "1 by 1" squares (Figure 55), and so

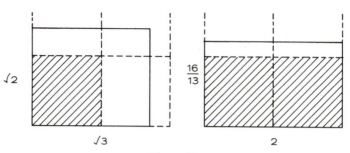

Figure 55

obtain the inequalities

$$1 < \text{area ABCD} < 4, \qquad 2 < \text{area A'B'C'D'} < 4.$$

Next we measure each rectangle using ".1 by .1" squares (Figure 56), and

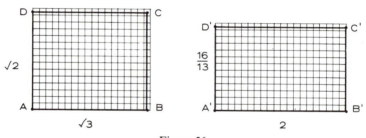

Figure 56

obtain the inequalities

$$14 \times 17 \times (.1)^2 < \text{area ABCD} < 15 \times 18 \times (.1)^2, \quad 12 \times 20 \times (.1)^2 < \text{area A'B'C'D'} < 13 \times 20 \times (.1)^2.$$

Next we measure each rectangle using ".01 by .01" squares, and so on. Continuing in this way we obtain a whole sequence of inequalities

| | |
|---|---|
| *1* < area ABCD < 4 | *2* < area A'B'C'D' < 4 |
| *2.38* < area ABCD < 2.70 | *2.4* < area A'B'C'D' < 2.6 |
| *2.4393* < area ABCD < 2.4708 | *2.46* < area A'B'C'D' < 2.48 |
| *2.449048* < area ABCD < 2.452195 | *2.460* < area A'B'C'D' < 2.462 |
| ... < area ABCD < | ... < area A'B'C'D' < ... |
| $\vdots$ | $\vdots$ |

When we examine these inequalities more carefully, we discover, in particular, an endless sequence of *lower* (or *inner*) approximations to the area of

ABCD: namely

$$a_0 = 1, \quad a_1 = 2.38, \quad a_2 = 2.4393, \quad a_3 = 2.449048, \quad a_4 = \ldots, \quad \ldots.$$

The terms of this sequence keep increasing; and since each $a_i$ is less than the area of ABCD, and (area ABCD) is less than 4, we know that each $a_i$ is less than the number $K = 4$ (though we could just as well have chosen $K = 2.70$, or $K = 2.4708$, or $K = 2.452195$, or ...). Hence by the fundamental property of real numbers (page 119), this endless sequence automatically pinpoints a real number $\alpha$, which must be less than or equal to $K$. Thus for each $i( = 0, 1, 2, \ldots)$ $\alpha$ must satisfy the inequalities

$$a_i \leq \alpha \leq K.$$

It is natural to assume that the number $\alpha$ pinpointed in this way by the sequence of inner approximations is what we mean by *the area of the rectangle ABCD*. For rectangles this is in fact quite correct,[5] and we shall simply take it for granted. If we then choose the particular value $K = 2.452195( = 1.733 \times 1.415)$ we see that

$$\text{area ABCD} = \alpha \leq K = 2.452195.$$

Similarly for the rectangle A'B'C'D' we obtain an endless sequence of *lower* (or *inner*) approximations

$$b_0 = 2, \quad b_1 = 2.4, \quad b_2 = 2.46, \quad b_3 = 2.460, \quad b_4 = \ldots, \quad \ldots.$$

The terms of this sequence also keep increasing[6]; and since each $b_i$ is less than the area of A'B'C'D', and (area A'B'C'D') is less than 4, we know that each $b_i$ is less than the number $L = 4$ (though we could just as well have chosen $L = 2.6$, or $L = 2.48$, or $L = 2.462$, or ...). Hence by the fundamental property of real numbers this endless sequence automatically pinpoints a real number $\beta$, which must be less than or equal to $L$; thus for each $i( = 0, 1, 2, \ldots)$ $\beta$ must satisfy the inequalities

$$b_i \leq \beta \leq L.$$

---

[5] You are invited to prove this in Exercise 6. The proof can be adapted to suit most of the shapes with which you are familiar. But for more general shapes in 2-dimensions it is simply *not true* that the number $\alpha$ pinpointed by the sequence of inner approximations is equal to the area (see Exercise 7(i)). *Even when the shape in question actually has an area*, the inner approximations worked out in the way we have suggested may get nowhere near it. The attempts by Jordan (1838–1922) and Borel (1871–1956) to make sense of this approach to *area* eventually helped Lebesgue (1875–1941) to introduce a much more general definition of *area* and of *integration*. But however area is defined it turns out that, if it is to have certain very simple properties (allowing us, for example, to cut up a shape and rearrange the pieces without changing its total area), then we even *have to abandon the idea that every shape automatically has an area* (see Exercise 7 (ii)).

[6] Though the fact that $b_2 = b_3 = 2.46$ shows that they do not always increase "strictly"—see Exercise 8.

Since we are measuring a rectangle, the number $\beta$ is equal to the area A′B′C′D′. If we choose the particular value $i = 2$, then we see that

$$2.46 = b_2 \le \beta = \text{area A′B′C′D′}.$$

By combining this with our previous inequality

$$\text{area ABCD} = \alpha \le 2.452195$$

we see that

$$\text{area ABCD} = \alpha < \beta = \text{area A′B′C′D′}.$$

This gives our first information about the ratio (area ABCD/area A′B′C′D′), namely

$$\frac{\alpha}{\beta} = \frac{\text{area ABCD}}{\text{area A′B′C′D′}} < 1.$$

But if we wish to calculate the ratio $\alpha/\beta$, then we must use an approach similar to that of Chapter II.11 (pages 129–130), where we obtained an inequality relating $\alpha/\beta$ and $a_i/b_i$: namely

$$\left| \frac{\alpha}{\beta} - \frac{a_i}{b_i} \right| \le \frac{\alpha + \beta}{\beta b_i} \cdot \frac{1}{10^i}.$$

In our case the apparently complicated right-hand side *is really very simple*, because

$$\alpha < 2.452195, \quad \beta < 2.462, \quad \text{and (when } i \ge 1) \quad \beta \cdot b_i > (2.4)^2,$$

so

$$\frac{\alpha + \beta}{\beta \cdot b_i} < \frac{2.452195 + 2.462}{(2.4)^2} < 1.$$

Therefore in our problem we know that, when $i \ge 1$,

$$\left| \frac{\alpha}{\beta} - \frac{a_i}{b_i} \right| < \frac{1}{10^i}.$$

This means that, though we do not know $\alpha$ and $\beta$ exactly, we can nevertheless calculate $\alpha/\beta$ to within $1/10^i$ by working out $a_i/b_i$ instead (and you are encouraged to do precisely this in Exercise 9).

We have applied the method of inner and outer approximations to the comparison of rectangles. Before leaving this approach we shall show how it can be used to prove the familiar *length × breadth* formula for the area of a rectangle.

Suppose that, instead of comparing the rectangle ABCD with the "16/13 by 2" rectangle A′B′C′D′ we were to compare it with a "1 by 1" square WXYZ. The endless sequence

$$c_0, \quad c_1, \quad c_2, \quad c_3, \quad c_4, \ldots$$

of *lower* (or *inner*) approximations for the square WXYZ is particularly simple. Inside WXYZ we can fit

precisely 1 "1 by 1" square,

precisely $10^2$ ".1 by .1" squares,

precisely $10^4$ ".01 by .01" squares,

$$\vdots$$

and so it goes on. Hence

$$c_0 = 1, \quad c_1 = 1, \quad c_2 = 1, \quad c_3 = 1, \quad c_4 = 1, \dots$$

and we scarcely need the fundamental property of real numbers to tell us that this endless sequence pinpoints the real number $\gamma = 1$. Nor is it a surprise to discover that $\gamma$ is equal to the area of the square WXYZ. If we now use the inequality from page 130 (with $\gamma$ and $c_i$ in place of $\beta$ and $b_i$) then we see that

$$\left| \frac{\alpha}{\gamma} - \frac{a_i}{c_i} \right| \leq \frac{\alpha + \gamma}{\gamma \cdot c_i} \cdot \frac{1}{10^i} \, .$$

But, since $\gamma = c_i = 1$, the left-hand side is equal to

$$|\alpha - a_i|$$

and the right-hand side is simply

$$(\alpha + 1) \cdot \frac{1}{10^i}.$$

Since $\alpha < 2.452195$, we conclude that

$$|\alpha - a_i| < \frac{1}{10^{i-1}}.$$

In other words, we can calculate the area $\alpha$ of the rectangle ABCD to within $1/10^{i-1}$ by simply working out the inner approximation $a_i$.

*But what exactly is $a_i$? And how do we work it out?* Remember that $a_i$ is precisely $1/(10^i)^2$ times the number of "$1/10^i$ by $1/10^i$" squares which fit inside the "$\sqrt{2}$ by $\sqrt{3}$" rectangle ABCD (Figure 57), and this is given by

$$a_i = l_i \times b_i$$

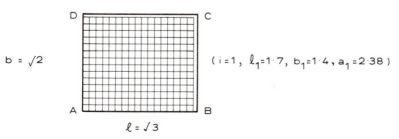

$$b = \sqrt{2}$$

$$( i = 1, \; l_1 = 1 \cdot 7, \; b_1 = 1 \cdot 4, \; a_1 = 2 \cdot 38 )$$

$$l = \sqrt{3}$$

Figure 57

where $l_i$ denotes the number given by the first $i$ decimal places of AB, and $b_i$ denotes the number given by the first $i$ decimal places of AD.[7] Now the sequence

$$l_0, \quad l_1, \quad l_2, \quad l_3, \quad l_4, \ldots$$

clearly pinpoints the length $l$ of AB, and the sequence

$$b_0, \quad b_1, \quad b_2, \quad b_3, \quad b_4, \ldots$$

clearly pinpoints the length $b$ of AD.

In Chapter II.11 we showed how, in this situation, the sequence of products

$$l_0 \times b_0, \quad l_1 \times b_1, \quad l_2 \times b_2, \quad l_3 \times b_3, \quad l_4 \times b_4, \ldots$$

*automatically pinpoints the product* $l \times b$; but since $l_i \times b_i = a_i$, this sequence of products is precisely the sequence

$$a_0, \quad a_1, \quad a_2, \quad a_3, \quad a_4, \ldots$$

*and we know already that this sequence pinpoints the real number* $\alpha(=\text{area ABCD})$. We have therefore proved that for our rectangle ABCD

$$l \times b = \alpha = \text{area ABCD}.$$

But this is precisely the *length* × *breadth* formula for the rectangle ABCD. Although ABCD here was a particular ("$\sqrt{2}$ by $\sqrt{3}$") rectangle, it should be fairly clear that exactly the same proof will work for any rectangle you may care to choose.

You should by now be convinced that it is possible to measure and to compare *rectangles* by approximating each rectangle inside and outside first using "1 by 1" squares, then using ".1 by .1" squares, then ".01 by .01" squares, and so on. This procedure therefore answers the first of our three questions entirely satisfactorily.

But what about measuring, or comparing, more general shapes—such as arbitrary polygons, or shapes with *curved* edges like the circle or ellipse? Although it is natural to try to use the same approach, we have already had a hint that the approach has its limitations—see Exercise 7. Nevertheless, if we are given an arbitrary shape X, let us *assume that it can be assigned an area* and try to measure this area using "1 by 1" squares (Figure 58).

This will give us crude *inner* and *outer* approximations ($a_0$ and $A_0$) to the area of the shape X:

$$a_0 \leq \text{area X} \leq A_0.$$

Next we measure using ".1 by .1" squares (Figure 59). This will give us slightly better inner and outer approximations ($a_1$ and $A_1$) and a second

---

[7] Thus

| | | | | |
|---|---|---|---|---|
| $l_0 = 1,$ | $l_1 = 1.7,$ | $l_2 = 1.73,$ | $l_3 = 1.732,$ | $\ldots$ |
| $b_0 = 1,$ | $b_1 = 1.4,$ | $b_2 = 1.41,$ | $b_3 = 1.414,$ | $\ldots$ |
| $a_0 = 1 \times 1,$ | $a_1 = 1.7 \times 1.4,$ | $a_2 = 1.73 \times 1.41,$ | $a_3 = 1.732 \times 1.414,$ | $\ldots$ |
| $= 1$ | $= 2.38$ | $= 2.4393$ | $= 2.449048$ | |

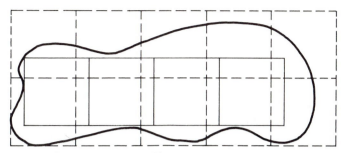

Figure 58

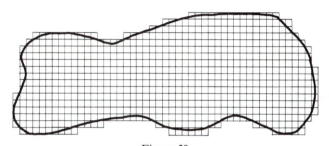

Figure 59

pair of inequalities

$$(a_0 \leq)a_1 \leq area \ X \leq A_1(\leq A_0).$$

Next we use ".01 by .01" squares and obtain still better inner and outer approximations ($a_2$ and $A_2$), and a third pair of inequalities

$$(a_0 \leq a_1 \leq)a_2 \leq area \ X \leq A_2( \leq A_1 \leq A_0).$$

And so it goes on.

We thus obtain *two* endless sequences. The first sequence is the endless sequence of inner approximations

$$a_0, \quad a_1, \quad a_2, \quad a_3, \quad a_4, \ldots .$$

The terms of this sequence keep increasing: since each $a_i$ is less than or equal to the area of the shape X, and since *area* X is less than or equal to $A_0$, we know that each $a_i$ is less than or equal to the number $K = A_0$ (though we could just as well have chosen $K = A_1$, or $K = A_2$, or ...). Hence by the fundamental property of real numbers, the endless sequence $a_0, a_1, a_2, \ldots$ of inner approximations automatically pinpoints some real number $\alpha$, which must be less than or equal to $K$. Thus for each $i( = 0, 1, 2, \ldots)$ $\alpha$ must satisfy the inequalities

$$a_i \leq \alpha \leq K.$$

And, since we could have chosen $K$ to be any one of the numbers $A_0$, $A_1$, $A_2, \ldots$, we see that $\alpha$ must in fact satisfy

$$a_0 \leq a_1 \leq a_2 \leq a_3 \leq \ldots \leq \alpha \leq \ldots \leq A_3 \leq A_2 \leq A_1 \leq A_0.$$

The second sequence is the endless sequence of outer approximations

$$A_0, \quad A_1, \quad A_2, \quad A_3, \quad A_4, \ldots .$$

The terms of this sequence keep decreasing; since each $A_i$ is greater than or equal to the area of the shape X, and since *area* X is greater than or equal to the number $a_0$, we know that each $A_i$ is greater than or equal to the number $k = a_0$ (though we could just as well have chosen $k = a_1$, or $k = a_2$, or ...). Hence, by the fundamental property of real numbers,[8] the endless sequence $A_0, A_1, A_2, \ldots$ of outer approximations automatically pinpoints some real number $A$, which must be greater than or equal to $k$. Thus for each $i(=0, 1, 2, \ldots)$ $A$ must satisfy the inequalities

$$k \leq A \leq A_i.$$

And, since we could have chosen $k$ to be any one of the numbers $a_0, a_1, a_2,$ ..., we see that $A$ must in fact satisfy

$$a_0 \leq a_1 \leq a_2 \leq a_3 \leq \ldots \leq A \leq \ldots \leq A_3 \leq A_2 \leq A_1 \leq A_0.$$

We can in fact do better than simply to squeeze $\alpha$ and $A$ between $a_0, a_1,$ $a_2, a_3, \ldots$ and $\ldots A_3, A_2, A_1, A_0$. For we can use purely arithmetical arguments (see Exercise 10) to prove that

$$\alpha \leq A.$$

You may feel it is intuitively obvious that the number $\alpha$ which is pinpointed by the *inner* approximations *cannot be greater than* the number $A$ which is pinpointed by the *outer* approximations; but remember that we are no longer willing to trust intuitively obvious statements unless they can be supported by strict mathematical proof. For example, it may seem equally obvious that the number $\alpha$ pinpointed by the inner approximations and the number $A$ pinpointed by the outer approximations are both equal to the area of the shape X, *but there is in fact no reason to expect that $\alpha = A$* (see for example, Exercise 7(i)). However, if the shape X is *sufficiently well-behaved* that the numbers $\alpha$ and $A$ pinpointed by the inner and outer approximations happen to be equal, then it is natural to call this number $\alpha(=A)$ *the area of the shape X*. It turns out that polygons are always sufficiently well-behaved in this sense, so our method of inner and outer approximations can be used to answer the second of our two questions; however, the details are rather messy and we shall not try to go through them here.

But there remain those awkward shapes for which

$$\alpha \text{ is strictly less than } A.$$

It would be very comforting if we could somehow *distinguish at a glance* between these apparently

*artificial,* or *unnatural,* specimens,

---

and what we believe to be

*ordinary, decent shapes!*

Unfortunately there is no sensible way of doing this.[9]

An alternative approach would be to refine our method of calculating inner and outer approximations in some way so as to produce inner (and outer) approximations $\bar{a}_i$ (resp. $\bar{A}_i$) which are *"better"* than those we obtained by using "$1/10^i$ by $1/10^i$" squares. The sequence

$$\bar{a}_1, \quad \bar{a}_2, \quad \bar{a}_3, \quad \bar{a}_4, \ldots$$

of *"improved"* inner approximations would then pinpoint a new real number $\bar{\alpha}$ $(\geq \alpha)$, and the sequence

$$\bar{A}_1, \quad \bar{A}_2, \quad \bar{A}_3, \quad \bar{A}_4, \ldots$$

of "improved" outer approximations would then pinpoint a new real number $\bar{A}$ $(\leq A)$. We might then hope to prove that we will always have

$$\bar{\alpha} = \bar{A}.$$

It is a fact of central importance in modern mathematics that our crude method of inner and outer approximations *can* be improved, and that as a result many shapes, which—like the awkward specimen in Exercise 7(i)(a)—could not previously be assigned an area, can in fact be shown to possess a well-defined area (see Exercise 7(iii)). However, *no matter how much the method may be refined, there will always exist other shapes which can still not be assigned an "area"*—though examples do not exactly spring to mind (see Exercise 7(ii)).

But however we proceed, it should by now be fairly clear that our intuitive concept of *area* needs to be much more carefully examined before we can safely use it to define other concepts—such as the definite integral of a function.

We shall leave the method of inner and outer approximations at this point and turn our attention to the alternative approach based on the idea of *dissection*. Our basic strategy here is the following:

given any shape X, we shall try to cut it up and reassemble all the pieces to make a *standard rectangle* X' — that is, a rectangle having one side of unit length (Figure 60).

---

[9] Unless, that is, you are so determined to exclude the awkward specimens that you are prepared to *define* a shape to be

*ordinary and decent*   when   $\alpha = A$

and to be

*artificial*   when   $\alpha < A$.

However, most people can see that such *monster-barring* tactics simply shirk the mathematical problem—see I. Lakatos: *Proofs and Refutations* (CUP, 1977), page 42.

Figure 60

If we then have to compare two shapes (X and Y), we simply reduce them to standard rectangles (X′ and Y′: Figure 61), and then compare these standard rectangles. The problem of comparing a "1 by α" standard

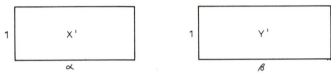

Figure 61

rectangle with a "1 by β" standard rectangle is precisely equivalent to the problem of comparing the two lengths α and β using the method of Chapter II.13 (see Exercise 1). The only real problem is therefore that of reducing the original shapes X and Y to standard rectangles X′ and Y′.

This dissection method is based on the *assumption* that if we dissect the shape X and rearrange the pieces so as to form a new shape X′, then the shapes X and X′ automatically have the same area.[10] Now we actually made use of this idea (without acknowledging it) in the method of inner and outer approximations. For when on page 176 we dissected the shape X into $a_0$ different "1 by 1" squares together with some remaining pieces which we shall lump together and simply call R, we quietly assumed
 (i) that, although the "1 by 1" squares were all in different positions, they all had the same area, and
(ii) that the area of the shape X is actually equal to

(the area of those $a_0$ squares) + (the area of R).

By appealing to another hidden assumption (namely that area R ≥ 0), we then deduced that

$$a_0 \leq \text{area } X \ (=a_0 + \text{area R}).$$

But let us now see how "dissection and rearrangement" can be used in the full-blooded dissection method.

---

[10] For ordinary length, area, and volume this assumption is valid as long as the pieces into which X is dissected are neither "too spiky," nor "too numerous." If, for example, one allows pieces which are so "spiky" that they cannot be assigned a volume, then it is theoretically possible to cut up the unit sphere $S$ into three congruent pieces $S_1, S_2, S_3$ which can then be reassembled to form a sphere twice as big as the original. If on the other hand one allows "too many" pieces, then one could cut up the unit interval [0, 1] into isolated points, and reassemble these separate points to make up the interval [0, 2] (see Exercise 11).

Suppose we wish to compare the same two rectangles as before (Figure 62), but this time using the dissection method rather than the method of

Figure 62

inner and outer approximations. The second rectangle Y can be easily reduced to a standard rectangle: first stand the rectangle Y on end (Figure 63), then cut along the line joining the midpoints M, N of the two sides of

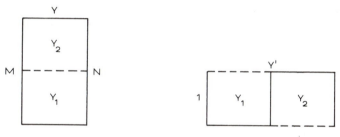

Figure 63

length 2, and place the two pieces side by side to make a "1 by 32/13" standard rectangle Y'.

We can scarcely expect every rectangle to be as easy to deal with as rectangle Y. But rectangle X is only slightly more awkward, and gives a clear idea of what one should do in the general case.

First of all, it turns out to be much easier to produce the segment of unit length which is to be one of the sides of our standard rectangle X', if we first reduce X to an "$a$ by $b$" rectangle with $a \leq 1$ and $b \geq 1$. To do this we repeatedly bisect the rectangle X and rearrange the pieces as in Figure 64.

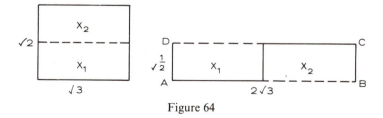

Figure 64

In general one may have to repeat this process several times, but in our particular example one such bisection is sufficient: we then land up with a "$1/\sqrt{2}$ by $2\sqrt{3}$" rectangle, (where $(1/\sqrt{2}) < 1$ and $2\sqrt{3} > 1$).

For our second step we draw a circular arc of radius 1 with centre at the corner A (Figure 65); this circular arc cuts the side AB in the point E, and

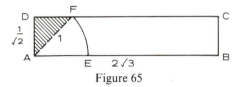

Figure 65

CD in the point $F$.[11] If we now cut along AF and shift the shaded triangle ADF to the other end—at BC (Figure 66), then we obtain a parallelogram

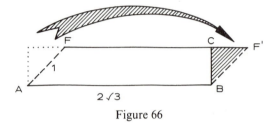

Figure 66

ABF′F with sides of length 1 and $2\sqrt{3}$.

Our third step will reduce this parallelogram to a standard rectangle X′ with AF as the side of unit length. The way we do this will be easier to follow if we draw the parallelogram so that the side AF runs vertically up the page (Figure 67). We shall now reduce the parallelogram ABF′F to the standard rectangle X′.

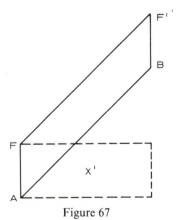

Figure 67

---

[11] However perverse it may seem, we should strictly explain *why* the circular arc automatically cuts the side CD. And though it certainly seems "obvious," the proof of the assertion that such lines necessarily cross depends on a geometrical version of the fundamental property of real numbers (though, admittedly, if one only wishes to prove the result for circles and straight lines then one *can* get away with much less than this—see J. Dieudonné: *Linear Algebra and Geometry* (Kershaw, 1969), Axiom (R15), page 18).

The line through F perpendicular to AF meets AB at the point G (Figure 68).

The line through G perpendicular to FG meets FF′ at the point H.

The line through H perpendicular to GH meets AB at the point I.

The line through I perpendicular to HI meets FF′ at the point J.

And finally, the line through J perpendicular to IJ meets BF′ at K.

If we now cut along FG, GH, HI, IJ, and JK, then the parallelogram ABF′F falls into six pieces $X_1, X_2, X_3, X_4, X_5, X_6$ and we can reassemble these pieces in a systematic way (Figure 69) to produce a "1 by $\sqrt{6}$" rectangle X′.[12]

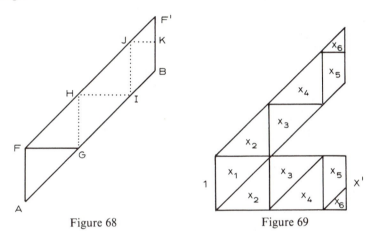

Figure 68                                            Figure 69

We have therefore reduced the problem of comparing the original rectangles X and Y to the much simpler problem of comparing the two standard rectangles X′ ("1 by $\sqrt{6}$") and Y′ ("1 by 32/13"). By Exercise 1, the ratio (area X′/area Y′) is equal to the ratio $\sqrt{6}/(32/13)$ of the "lengths" of the two rectangles X′ and Y′. Hence

$$\frac{\text{area X}}{\text{area Y}} = \frac{\text{area X}'}{\text{area Y}'} = \frac{\sqrt{6}}{32/13} \left( = \sqrt{\frac{1014}{1024}} \right).$$

We may then calculate both 1014/1024 and its square root as accurately as we please.

Any other pair of rectangles X, Y can be compared in exactly the same way.

But where do we go from here? Rectangles may be a necessary starting point, but they aren't exactly very exciting. However though it might at first

---

[12] X′ is obviously a "1 by *something*" (say a "1 by *x*") rectangle. But since we are offering the dissection method as *an alternative to* the method of inner and outer approximations we should not, strictly speaking, use the *"length × breadth"* formula to show that X′ is a "1 by $\sqrt{6}$" rectangle (that is, we should *not* simply say $\sqrt{2} \times \sqrt{3} = $ area X = area X′ = 1 × *x*). In Exercise 12 we show how to use similar triangles to find *x*.

sight appear that the problem of reducing a given rectangle X to a standard rectangle X′ was rather special, it is in fact quite easy to extend what we have done for rectangles first to

*parallelograms*

then to

*triangles*

and finally to

*general polygons.*

We shall not describe the process in enormous detail, but the main ideas should be fairly clear.

First of all, by cutting once and rearranging the two pieces we can reduce any parallelogram ABCD to a rectangle ABEE′ with one side equal to AB (Figure 70). We can then reduce this rectangle to a standard ("1 by *x*") rectangle as we did before. Thus *every parallelogram can be reduced to a standard rectangle.*

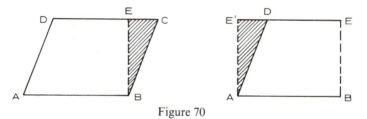

Figure 70

Secondly, by cutting once and rearranging the two pieces, any triangle ABC can be reduced to a parallelogram ABM′M (Figure 71). The parallelogram ABM′M can then be reduced to a standard rectangle as before. Thus *every triangle can be reduced to a standard rectangle.*

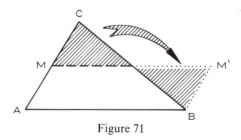

Figure 71

Thirdly, any quadrilateral can be dissected with a single cut into two triangles X and Y (Figure 72). But we have just seen how these two triangles X and Y can be reduced to standard rectangles X′ ("1 by *x*") and Y′ ("1 by *y*"). These standard rectangles X′ and Y′ can then be put together (Figure

Figure 72

73) to make a "1 by $(x + y)$" standard rectangle which must be equal in area to the quadrilateral we started with.

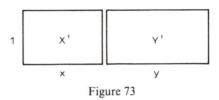

Figure 73

Similarly, any pentagon can be dissected with a single cut into a triangle X and a quadrilateral Y (Figure 74)—see Exercise 13(i). But we have seen how the triangle X can be reduced to a standard ("1 by $x$") rectangle X', and also how the quadrilateral Y can be reduced to a standard ("1 by $y$") rectangle Y'. These two standard rectangles can then be put together to make a "1 by $(x + y)$" standard rectangle, which must be equal in area to the pentagon we started with.

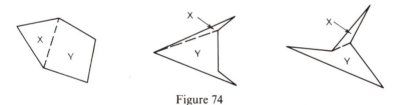

Figure 74

In general, once we know how to reduce any triangle (that is, any polygon with 3 sides), and any polygon with $n$ sides to a standard rectangle, then we can reduce any polygon with $n + 1$ sides to a standard rectangle (Exercise 13(ii)). *Since we have proved that every triangle can in fact be reduced to a standard rectangle*, it follows (by induction) that every polygon can be reduced to a standard rectangle. We can therefore compare any two polygons by reducing each one in turn to a standard rectangle, and then comparing the standard rectangles.

But what about 2-dimensional shapes with curved edges like those in Figure 75? One cannot in general expect to find a way of dissecting shapes like these in such a way that the pieces can be reassembled to make a

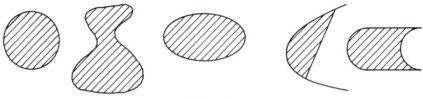

Figure 75

rectangle (though this can in fact be done with the last shape in Figure 75—but that is only because the "curviness" of its edge consists simply of a semicircular lump and a semicircular dent of exactly the same size). So how should we begin? How, for example, can we *prove* the familiar formula for the area of a circular disc?

The whole question of finding the area of curved shapes was extensively studied by the ancient Greeks.[13] They made very skilful use of dissection; but they also used a procedure related to our method of inner and outer approximations. One of the reasons why their approach differed from ours deserves to be mentioned explicitly[14]—this chapter is, after all, meant to convince you that a strict mathematical treatment of area is more complicated than you might have suspected. We have simply glossed over the fact that, though *squares* fit in nice tidy rows inside a *rectangle*, they are really not much use for actually measuring the areas of more general shapes. For example, it is not at all easy to decide the exact number of ".1 by .1" squares which can be fitted inside a circle of radius 1 (Exercise 14). Thus in practice for each new shape X we have to find some method of approximation which suits the particular shape X we happen to be working with. This can require considerable ingenuity. But if the approach requires such ingenuity for every special case, it surely loses all claim to being a truly *general method!* Moreover the way we decide to approximate the shape X is now *highly subjective*, so we have the theoretical problem of deciding whether two different methods of approximation always give the same answer.

Thus in spite of some brilliant individual discoveries—such as Archimedes' deservedly famous proof that the area of a segment of a parabola is

---

[13] They formulated the problem of finding the area of a curved shape X in the following way:

*find a square whose area is equal to the area of X.*

The dominance of the Latin language in Christian Europe meant that the whole subject of "calculating areas of curved shapes" came to be known as

*the* quadrature *of curves*      (Latin: quadratus = square).

[14] A second reason is of considerable historical interest. As it stands, our method of inner and outer approximations requires, in principle, that we generate an *endless* sequence of inner (and outer) approximations. As we have mentioned before, Greek mathematics at its height (ca. 300 BC) recognised that such endless infinite processes had no place in mathematics as they understood it. If however the answer could be guessed in advance—perhaps by secretly using an "illicit" infinite process—they were able to use the extremely delicate *method of exhaustion*, which is reminiscent in many ways of the "$\varepsilon$-$\delta$" proofs you would meet in an analysis course.

equal to 4/3 of the area of the largest triangle it contains (Figure 76)—it seems fair to say that the Greeks had no effective general method for calculating the areas of curved shapes.

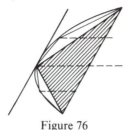

Figure 76

The sixteenth and seventeenth centuries witnessed a massive resurgence of interest in the quadrature problem, and many new ingenious ways were found to calculate the areas of particular curved shapes. *But the simple truth remains that it was the invention of the integral calculus that provided the first general method for calculating areas of curved shapes.*

But if integration represents our *only* means of actually calculating most areas, is it not a circular argument to *define* the integral of $f(x)$ in terms of the area under the curve $y = f(x)$? If we wish to define the integral of $f(x)$ in this way we must presumably understand areas of curved shapes *first*, whereas the facts seem to suggest that our understanding of areas of curved shapes in general is in some sense dependent on the integral itself.

In this chapter we have examined the intuitive notion of "the area of a 2-dimensional shape" only in sufficient detail to indicate

(i) the sorts of difficulties that arise when we eventually come to consider shapes with curved edges, and

(ii) the logical difficulty of *defining* the integral of a function in terms of "the area under its graph."

Our aim was simply to suggest some of the grounds for deliberately excluding geometrical notions in the reconstruction of the calculus—a feature which is one of the hallmarks of the 1870 version of the calculus. The full story is much more complex: in particular the most significant developments concerning area and integration have taken place since 1870—but the reader interested in these developments can do no better than consult the detailed, but eminently readable, book by Thomas Hawkins: *Lebesgue's Theory of Integration* (most recently published by the Chelsea Publishing Co. in 1975), or the chapter by Hawkins in I. Grattan Guinness (Ed.): *From the Calculus to Set Theory*, 1630–1910 (Duckworth, 1980).

EXERCISES

1. Let X be a "1 by $\alpha$" rectangle with side AB of length $\alpha$, and Y be a "1 by $\beta$" rectangle with side CD of length $\beta$ (Figure 77). The procedure of Chapter II.13

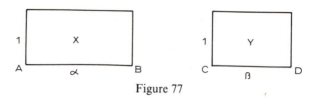

Figure 77

can be used to obtain the continued fraction expansion for the ratio AB/CD:

$$\frac{AB}{CD} = [a_0; a_1, a_2, a_3, \ldots].$$

We can also apply the same procedure to the rectangles X and Y to obtain the continued fraction expansion for the ratio X/Y:

$$\frac{X}{Y} = [b_0; b_1, b_2, b_3, \ldots]$$

(that is, first Y fits precisely $b_0$ times into X, leaving the remainder X' (Figure 78); then X' fits precisely $b_1$ times into Y, leaving the remainder Y'; and so on).

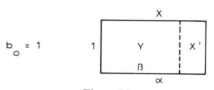

Figure 78

(i) Show that $b_0 = a_0$.
(ii) Prove, by induction on $n$, that $a_n = b_n$ for every $n$. [Hint: Assume that $a_i = b_i$ for each $i = 0, 1, \ldots, k$ and then prove that $a_{k+1} = b_{k+1}$. Then use part (i).]
(iii) Hence conclude that

$$\frac{X}{Y} = \frac{AB}{CD} = \frac{\alpha}{\beta}.$$

2. Consider the "proof" of the formula for the area of a circular disc of radius $r$ presented in Figure 79. Usually it is simply *asserted* that, as the number $N$ of segments increases, the *rearranged shape* becomes more or less indistinguishable from a "$\pi r$ by $r$" rectangle.

(i) Draw the smallest rectangle which completely covers each rearranged shape, and calculate the lengths of its sides (in terms of $N$).
    Draw the largest rectangle which fits inside each rearranged shape, and calculate the lengths of its sides in terms of $N$.

(ii) As $N$ increases, what happens to the lengths of the sides, and to the areas, of these *inner* and *outer* approximating rectangles?
[The *outer* rectangle is "$r(2 - \cos(\pi/N))$ by $(N + 1)r \sin(\pi/N)$" and the *inner* rectangle is "$r \cos(\pi/N)$ by $(N - 1)r \sin(\pi/N)$". Use a calculator to evaluate the sides and the areas of these rectangles when $r = 1$, for $N = 3, 4, 5, \ldots, 16, \ldots$.

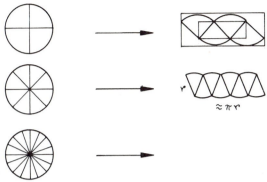

Figure 79

Then try to find a *proof* which shows

(a) that the areas $a'_3$, $a'_4$, $a'_5$, ... of the inner approximating rectangles, and the areas $A'_3$, $A'_4$, $A'_5$, ... of the outer approximating rectangles actually pin-point the same real number; and

(b) that this real number is equal to $\pi r^2$.]

(iii) The usual argument (that the rearranged shape is like a "$\pi r$ by $r$" rectangle) assumes that the circumference of a circle of radius $r$ is equal to $2\pi r$. How can one *prove* this familiar fact?

(iv) Show that although the sequence of wiggly lines in Figure 80 gets closer and closer to the line segment AB, the total length of each wiggly line remains $\sqrt{2}$ times as long as the segment AB.

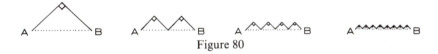

Figure 80

(v) Let us assume the familiar formula $2\pi r$ for the circumference of a circle of radius $r$. How can we be sure that the length of the straight segment AB corresponding to the wiggly side of the "near-rectangle" in Figure 81 gets closer and closer to $\pi r$?

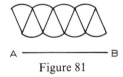

A ———————— B

Figure 81

3.  (i)   If $\alpha$ is an irrational number, show that $\alpha/2$ is also irrational.

(ii)  If $\beta$ is an irrational number, show that $\sqrt{\beta}$ is also irrational.

(iii) *Assume* that $\pi^2$ is irrational, and show that $\pi/\sqrt{2}$ is also irrational.

(iv) Prove that $\pi^2$ is irrational. [Hint: The proof is not too hard, but it is probably more intricate than proofs you have met before. The separate steps are specified below: you should try to complete one step at a time, pausing every so often—perhaps after Step 4 and Step 8—to see how things are going. The complete proof is written out in Chapter 16 of M. Spivak: *Calculus* (Addison-Wesley, (1973); 2nd edition Publish or Perish Inc., (1980)).

*Step 1 :* Define

$$f_n(x) = \frac{x^n(1-x)^n}{n!}.$$

Show

$$0 < f_n(x) < \frac{1}{n!} \quad \text{for} \quad 0 < x < 1.$$

*Step 2 :* Expand $f_n(x)$ as a polynomial and show

$$f_n(x) = \frac{1}{n!} \sum_{i=n}^{2n} c_i x^i$$

where each $c_i$ is a whole number.

*Step 3 :* If $f_n^{(k)}$ denotes the $k^{\text{th}}$ derivative of the function $f_n$, show that

$$f_n^{(k)}(0) = 0 \quad \text{for} \quad k < n \quad \text{or} \quad k > 2n,$$

and

$$f_n^{(k)}(0) = \frac{k!}{n!} c_k \quad \text{for} \quad n \le k \le 2n.$$

Conclude that $f_n^{(k)}(0)$ *is a whole number* for each choice of $k$.

*Step 4 :* Show that $f_n(x) = f_n(1-x)$. Deduce that $f_n^{(k)}(x) = (-1)^k f_n^{(k)}(1-x)$, and hence show that $f_n^{(k)}(1)$ *is a whole number* of each choice of $k$.

*Step 5 :* Suppose $\pi^2 = a/b$ where $a$ and $b$ are positive whole numbers. Define $G(x) = b^n[\pi^{2n} f_n(x) - \pi^{2n-2} f_n^{(2)}(x) + \pi^{2n-4} f_n^{(4)}(x) - \ldots + (-1)^n f_n^{(2n)}(x)]$. Show $b^n \cdot \pi^{2n-2k} = a^{n-k} \cdot b^k$ is a whole number; deduce that $G(0)$ and $G(1)$ are whole numbers.

*Step 6 :* Differentiate $G$ twice to obtain

$$G^{(2)}(x) = b^n[\pi^{2n} f_n^{(2)}(x) - \ldots + (-1)^n f_n^{(2n+2)}(x)].$$

Then add $\pi^2 G(x)$ and simplify to obtain

$$G^{(2)}(x) + \pi^2 G(x) = b^n \cdot \pi^{2n+2} \cdot f_n(x) = \pi^2 \cdot a^n \cdot f_n(x).$$

*Step 7 :* Define $H(x) = G^{(1)}(x) \sin \pi x - \pi G(x) \cos \pi x$, and show that

$$H^{(1)}(x) = \pi^2 \cdot a^n \cdot f_n(x) \cdot \sin \pi x.$$

*Step 8 :* $\pi^2 \displaystyle\int_0^1 a^n f_n(x) \sin \pi x \, dx = \int_0^1 H^{(1)}(x) \, dx$

$$= H(1) - H(0) = \pi(G(1) + G(0)).$$

Conclude that $\pi \int_0^1 a^n f_n(x) \sin \pi x \, dx$ *is a whole number.*

*Step 9 :* We *assumed* that $\pi^2 = a/b$; if we can show that

$$0 < \pi \int_0^1 a^n f_n(x) \sin \pi x \, dx < 1$$

then we will have a contradiction, and so may assert that

$$\pi^2 \text{ is irrational.}$$

The contradiction is reached as follows:

(a) $0 < \pi a^n f_n(x)\sin \pi x < (\pi a^n/n!)$ by (i).
(b) Hence $0 < \pi \int_0^1 a^n f_n(x)\sin \pi x \, dx < (\pi a^n/n!)$.
(c) But given any fixed positive number $a$, we can choose $n$ to make $0 < (\pi a^n/n!) < 1$.]

4. (i) Suppose we have been given a curve $\gamma$ and a point $P$ on that curve (Figure 82). *What exactly does it mean to say that a straight line $l$ is tangent to the curve at the point $P$?* We need a precise definition

(a) so that we can recognise what is, and what is not, a tangent, and
(b) so that we can prove results about tangents in general.

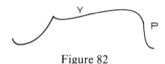

Figure 82

Well, how about the following?

**Definition 1:** "*$l$ is tangent to $\gamma$ at $P$*" means simply that "*$l$ meets the curve $\gamma$ at $P$ only.*"

But how can we square Definition 1 with the examples in Figure 83? We certainly never meant to exclude Example 1B, so let's try again.

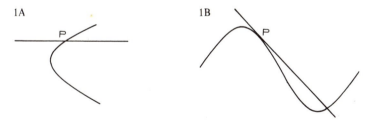

Figure 83.   Example 1A fits Definition 1; Example 1B does not fit Definition 1.

**Definition 2:** "*$l$ is tangent to $\gamma$ at $P$*" means simply that "*$l$ meets the curve $\gamma$ at $P$ but meets it nowhere else 'near' $P$.*"

But how can we square Definition 2 with the examples in Figure 84? Lets put Example 2B aside for the moment and try to exclude Example 2A. How about

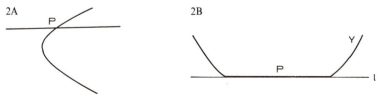

Figure 84.   Example 2A fits Definition 2; Example 2B does not fit Definition 2.

**Definition 3:** *"l is tangent to γ at P" means simply that "l meets the curve γ at P, but does not 'cross' the curve at P."*

The examples in Figure 85 show that Definition 3 is still inadequate. Thus none of these "definitions" captures what we mean by

*"l is tangent to γ at P.'*

Figure 85.   Example 3A fits Definition 3; Example 3B does not fit Definition 3.

You should certainly try to improve on Definitions 1, 2, and 3, but do not expect complete success to come easily (—you may even begin to feel some sympathy with Samuel Butler, who wrote in his Notebooks: "Definitions are a kind of scratching and generally leave a sore place more sore than it was before").

For the purely arithmetical formulation of the idea of a tangent to a curve, it is convenient to restrict attention (at least in the first instance) to curves which arise as the graph $y = f(x)$ of some function $f(x)$ (Figure 86). Part of the difficulty we had in Definitions 1, 2, and 3 can be avoided by first splitting the original question into two parts:

(a) *First (and most importantly) explain what it means to say that the curve $y = f(x)$ actually possesses a tangent at the point P.*
(b) Then explain how to recognise the tangent line if it exists.

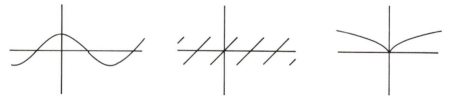

Figure 86

It may be worth simply stating one purely arithmetical answer to the first of these questions, even though we shall make no attempt in this book to explain the curious form it takes:

> *The curve $y = f(x)$ has a tangent at the point $P = (x_0, f(x_0))$ precisely when there exists a number s with the following property:*
> *for each number $\varepsilon > 0$,*
> *we must be able to find another number $h > 0$, such that*
> *whenever x is a number satisfying the inequalities $x_0 - h < x < x_0 + h$,*
> *then x automatically satisfies the inequalities $s - \varepsilon < (f(x) - f(x_0))/h < s + \varepsilon$.*

(ii) Let us stick to curves which arise as the graph $y = f(x)$ of some function $f(x)$. *What exactly does it mean to say that such a curve is continuous?*

To familiarise yourself with the nature of the problem here, try to formulate a rough *Definition 1*; then criticise your definition by producing examples—as we did for tangents in part (i); then improve on your first effort by formulating a *Definition 2*. Repeat this process until you either run out of ideas, or succeed in obtaining a clear definition which seems to withstand all criticism!

(The purpose of this exercise is not just to find a valid definition—indeed it would be a remarkable achievement if you managed to find one. What is important is that you should actually work through the various stages of

*rough formulation,*
*criticism,*
*improvement,*

in order to get some feeling for the genuine difficulty of producing a precise definition. The whole exercise is likely to be far more fruitful if it is tackled by small groups of students working together.)

The purely arithmetic formulation of the notion of continuity again splits the problem into two parts. Though one naturally thinks of *continuity* as a property of *the curve as a whole*, if we turn the idea on its head for a moment, and think about *discontinuity* (Figure 87), then one gets a fairly strong impression that

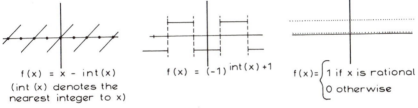

$f(x) = x - int(x)$
(int (x) denotes the
nearest integer to x)

$f(x) = (-1)^{int(x)+1}$

$f(x) = \begin{cases} 1 \text{ if } x \text{ is rational} \\ 0 \text{ otherwise} \end{cases}$

Figure 87

*discontinuities occur for particular values of x*—though this is not so striking in the last of our three examples since that "curve" happens to be discontinuous at *every* point. Thus

> the first, and most important, step must be to explain what it means to say that the curve $y = f(x)$ is discontinuous (or continuous) at a particular point $(x_0, f(x_0))$.

As in part (i), it may be worth simply stating one purely arithmetical solution to the problem of defining "continuity at a point" even though we shall make no attempt to convince you that it is a useful definition which really captures what we mean intuitively by "continuity":

> The curve $y = f(x)$ is continuous at the point $(x_0, f(x_0))$ precisely when
> for each number $\varepsilon > 0$,
> we are able to find another number $h > 0$, which has the property that
> whenever x is a number satisfying the inequalities $x_0 - h < x < x_0 + h$,
> then x automatically satisfies the inequalities $f(x_0) - \varepsilon < f(x) < f(x_0) + \varepsilon$.

(iii) In parts (i) and (ii) we rather shirked the question

*What is a curve?*

by discussing only those curves which arise as the graph $y = f(x)$ of some function $f(x)$. We therefore reduced the *geometric* question—about *curves*—to a (more or less) *arithmetic* question—about *functions*. We should therefore, strictly speaking, now answer the question

*What is a function?*

This question is neither as silly, nor as easy to answer, as one might think. But we have already overstayed our welcome in this particular Exercise, and so must postpone discussion of this question until Part IV.

5.   (i) Show that $\sqrt{3/2}$ is irrational.
   (ii) Show that $\sqrt{2/(16/13)}$ is irrational.
   (iii) Show that if $x \neq 0$ is rational and $y$ is irrational, then $x \cdot y$ is irrational.
   (iv) Find irrationals $x$, $y$, $z$ such that $x \cdot y$ is rational and $x \cdot z$ is irrational.

6. Let ABCD be an "$l$ by $b$" rectangle. We shall use the same notation as was used in the text:

   $l_i$ denotes the number formed by taking the first $i$ decimal places of $l$;
   $b_i$ denotes the number formed by taking the first $i$ decimal places of $b$;
   $a_i$ denotes the "area" of "$1/10^i$ by $1/10^i$" squares which fit inside ABCD;
   $A_i$ denotes the "area" of "$1/10^i$ by $1/10^i$" squares needed to cover ABCD;
   $\alpha$ denotes the real number which is pinpointed by the increasing sequence

$$a_0, \quad a_1, \quad a_2, \ldots ;$$

$A$ denotes the real number which is pinpointed by the decreasing sequence

$$A_0, \quad A_1, \quad A_2, \ldots .$$

Use the fact (page 175) that

$$a_i = l_i \times b_i$$

to show that

$$A_i = a_i + \frac{10^i(l_i + b_i) + 1}{(10^i)^2}.$$

Deduce that

$$\alpha = A.$$

7. In the text we have used the word "shape" without being very specific as to what is, and what is not, a "*shape*." Perhaps you imagined that the word was supposed to refer only to familiar shapes, like circular discs, rectangles, ellipses, and so on. If this is what you had assumed, then you may wish to complain that the examples in this exercise *are not really shapes at all*. But if this is how you react, then, to be fair, you should explain exactly what *you* mean by a *shape in 2-dimensions*. (Even if this is not how you react, then it is still worth trying to explain, in the spirit of Exercise 4, what you mean by a "shape" in 2-dimensions.)

   In practise there seems to be no obvious way of distinguishing between "nice,

well-behaved shapes" and "horrid, spiky monsters," so we shall simply assume
here that

"*2-dimensional shape*"   means precisely   "*some subset of the plane*,"

and that

"*1-dimensional shape*"   means precisely   "*some subset of the real line*".

(i) (a) Let X be the subset of the plane which consists precisely of the line
segments joining $(0, 0)$ to each of the points $(r, \pm 1)$, $(\pm 1, r)$, where $r$ runs
through all the rational numbers between $-1$ and $+1$ inclusive (Figure 88).

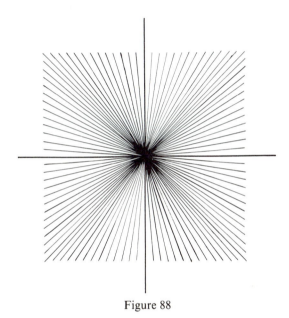

Figure 88

Show that for the shape X

$$0 = a_0 = a_1 = a_2 = \ldots \quad \text{and} \quad 4 = A_0 = A_1 = A_2 = \ldots .$$

Hence show that

$$\alpha = 0 < 4 = A.$$

(b) Let Y be the subset of the plane defined in the following way: Let $Y_0$ be a
"1 by 1" square (Figure 89). Let $Y_1$ be obtained from $Y_0$ by removing the four
"1/3 by 1/3" squares in the middle of each edge (Figure 90). Let $Y_2$ be obtained

Figure 89                Figure 90                Figure 91

from $Y_1$ by removing *from each of the five remaining pieces* the four "1/9 by 1/9" squares in the middle of each edge (Figure 91). $Y_3$ is obtained from $Y_2$ in a similar way; then $Y_4$ from $Y_3$; and so on. Finally Y is defined to consist of precisely *those points which belong to every* $Y_i (i = 0, 1, 2, \ldots)$.

To approximate Y from inside and out it is more convenient to define $a_i$ and $A_i$ in terms of "$1/3^i$ by $1/3^i$" squares (rather than "$1/10^i$ by $1/10^i$" squares). Show that

$$a_0 = a_1 = 0 \quad \text{and} \quad A_0 = 1, \quad A_1 = \frac{5}{9}.$$

Find expressions for the general terms $a_n$ and $A_n$. Hence find the area of the shape Y.

(ii) In the text we have asserted that there exist shapes which cannot be assigned an area. There are indeed many many such shapes, but they tend to be rather complicated. We shall therefore begin with a slightly simpler problem, and shall describe a *1-dimensional shape* X *which cannot be assigned a length*. Afterwards we shall indicate how the same idea can be used to describe a *2-dimensional shape which cannot be assigned an area*. We proceed in two stages:

*first* describe the set X;
*then* prove that X cannot possibly be assigned a length.

For the second stage we use the method of *proof by contradiction*: in other words, we assume that X can be assigned a length, and then appeal to *properties of lengths* to obtain a contradiction.

Although this second stage is much easier than the first stage, the properties of lengths which we shall appeal to in the second stage have a direct influence on everything we do in the first stage. So let us start by mentioning the four properties of length which we shall use:

(1) the length of an ordinary interval $(a, b)$ is equal to $b - a$;
(2) If Y is a subset of Z, then the length of Y is less than or equal to the length of Z;
(3) if Y is any set, and $Y + r$ is obtained by translating Y *bodily* through a distance $r$—to the right if $r$ is positive, and to the left if $r$ is negative—then the length of Y is equal to the length of $Y + r$;
(4) if $Y_1, Y_2, Y_3, \ldots$ is a (possibly endless) sequence of sets, no two of which overlap, and each of which has a length, then

$$\text{length}\left( \bigcup_{i=1}^{\infty} Y_i \right) = \sum_{i=1}^{\infty} \text{length}(Y_i).$$

We shall now describe the set X.

(a) Introduce a relation $\sim$ on the set of real numbers by defining

"$x \sim y$" precisely when "$x - y$ is a rational number".

Show that $x \sim x$ for each real number $x$;

if $x \sim y$, then $y \sim x$;
if $x \sim y$ and $y \sim z$, then $x \sim z$.

The relation $\sim$ is therefore an equivalence relation, and so partitions the set of real numbers into equivalence classes. Let $[x]$ denote the equivalence class which contains the number $x$. Describe the class $[0]$, and show that $[\sqrt{2}] \neq [\sqrt{3}]$.

(b) Show that each equivalence class contains at least one number between 0 and 1.

The set X consists of *just one* number from each equivalence class, these numbers being chosen so as to lie between 0 and 1.

(c) If $r$ is any rational number, then

$$X + r = \{x + r : x \in X\}$$

is the set obtained from X by simply translating it bodily through a distance $r$ (to the right if $r$ is positive, and to the left if $r$ is negative). Show that, if $r \neq s$, then $X + r$ and $X + s$ have no element in common.

(d) We must now make use of the following fact: *It is possible to include all the rational numbers which lie between $-1$ and $+1$ in a single endless sequence*

$$r_1, \quad r_2, \quad r_3, \ldots.$$

(The order in which the rationals occur in such a sequence has nothing to do with the natural order of the rationals according to size. Here is the beginning of one such endless sequence containing all the rationals between $-1$ and $+1$: can you see how it is meant to continue?

$$\frac{0}{1}, \frac{1}{1}, \frac{-1}{1}, \frac{1}{2}, \frac{-1}{2}, \frac{1}{3}, \frac{-1}{3}, \frac{1}{4}, \frac{-1}{4}, \frac{2}{3}, \frac{-2}{3}, \frac{1}{5}, \frac{-1}{5}, \frac{1}{6}, \ldots \Bigg)$$

Show that

$$(0, 1) \subseteq \bigcup_{i=1}^{\infty} (X + r_i) \subseteq (-1, 2).$$

(e) *Now assume that X can be assigned a length : length(X). Show that*

$$\text{length}(X + r_i) = \text{length}(X), \quad \text{for each } i = 1, 2, \ldots.$$

Show that

$$\text{length}\left( \bigcup_{i=1}^{\infty} (X + r_i) \right) = \sum_{i=1}^{\infty} \text{length}(X + r_i) = \sum_{i=1}^{\infty} \text{length}(X),$$

and that

$$1 \leq \text{length}\left( \bigcup_{i=1}^{\infty} (X + r_i) \right) \leq 3.$$

Conclude that the sum of infinitely many numbers (length(X + $r_i$)), all equal to length(X), must produce a number between 1 and 3. Hence

*length(X) cannot be zero*

(remember that the value of an endless sum is precisely the number pinpointed by the corresponding endless sequence of finite sums), and

*length(X) cannot be strictly positive*

(if length(X) = $\varepsilon$ > 0, then choosing a whole number $N$ such that $1/N < \varepsilon$ leads to the conclusion that $\sum_{i=1}^{\infty}$ length(X) $> \sum_{i=1}^{3N}$ length(X) $> 3$).

*We conclude from all this that our set X cannot in fact be assigned a length.*

To obtain an analogous set X in 2-dimensions, introduce a relation $\sim$ on $R^2$ by defining

"$(x_1, x_2) \sim (y_1, y_2)$" precisely when "$x_1 - y_1$ and $x_2 - y_2$ are rational."

Show that this is an equivalence relation. Then define X to consist of precisely one ordered pair $(x_1, x_2)$ from each equivalence class, these being chosen so as to lie inside the unit square with corners at (0, 0), (1, 0), (1, 1), (0, 1). Then proceed more or less as before.

(iii) Observe that the method of inner and outer approximations in the text only tries to fit *finitely many* squares inside or over each shape: when approximating a shape X, we fitted $a_0$ "1 by 1" squares completely inside X, and $A_0$ "1 by 1" squares completely covering X, and then deduced that

$$a_0 \leq \text{area X} \leq A_0;$$

subsequent steps also involved only *finitely many* squares. We then focussed our attention on the real numbers $\alpha$ and $A$ pinpointed by these two sequences of inner and outer approximations. Unfortunately, when we applied this method to the shape in part (i)(a), we discovered that

$$0 = \alpha < A = 4.$$

However, the fourth property of *length* which we stated in part (ii) constitutes a radically new approach: for it clearly admits the novel idea that the set

$$Y = \bigcup_{i=1}^{\infty} Y_i$$

may be dissected into an *endless* sequence of (disjoint) pieces

$$Y_1, \quad Y_2, \quad Y_3, \ldots$$

and asserts that the length of Y is then equal to the *endless* sum of the lengths of the separate pieces

$$\text{length(Y)} = \sum_{i=1}^{\infty} \text{length}(Y_i).$$

If we now allow ourselves the luxury of covering the shape in part (i)(a) by an *endless* sequence of pieces (and it must be stressed that this constitutes a radically new departure), then we obtain a refined sequence of outer approximations

$$\bar{A}_1, \quad \bar{A}_2, \quad \bar{A}_3, \ldots$$

which actually pinpoints the number $\bar{A} = 0$. Once we accept this new approach, it becomes natural to assign the area 0 to the shape in part (i)(a). *We shall now outline the proof that $\bar{A} = 0$.*

(a) The points $P$ of the form $(\pm 1, r)$ and $(r, \pm 1)$, where $r$ is any rational number between $-1$ and $+1$, can be included in a single endless sequence

$$P_1, \quad P_2, \quad P_3, \ldots .$$

(b) Let $N$ be any positive whole number. If 0 denotes the origin, show that the segment $OP_i$ can be completely covered by a triangle $T_{N,i}$ of area at most $1/(N \cdot 2^{i+1})$.

Conclude that the shape can be completely covered by the endless sequence of triangles

$$T_{N,1}, \quad T_{N,2}, \quad T_{N,3}, \ldots$$

whose total area $\bar{A}_N$ is less than $1/N$.

(c) This sequence of outer approximations

$$\bar{A}_1, \quad \bar{A}_2, \quad \bar{A}_3, \ldots$$

clearly pinpoints the number $\bar{A} = 0$.

8. Let ABCD be a rectangle, and $a_0, a_1, a_2, a_3, \ldots$ the increasing sequence of inner approximations.

(i) If the lengths of AB and AD can both be expressed as decimal fractions—say

$$\text{AB} = \frac{c}{10^n}, \qquad \text{AD} = \frac{d}{10^n},$$

(where $c$ and $d$ are both positive whole numbers), show that

$$a_n = a_{n+1} = a_{n+2} = \ldots = \text{area ABCD}.$$

(ii) Conversely, if all the $a_i$'s, from a certain point on, happen to be equal—say

$$a_i = a_{i+1} = a_{i+2} = \ldots,$$

show that the lengths of the sides AB and AD can both be expressed as decimal fractions of the form

$$\text{AB} = \frac{c}{10^i}, \qquad \text{AD} = \frac{d}{10^i}.$$

9. (i) In the text, ABCD is a "$\sqrt{2}$ by $\sqrt{3}$" rectangle, and $a_0, a_1, a_2, a_3, \ldots$ is the corresponding sequence of inner approximations which pinpoints the number $\alpha = \text{area ABCD}$.

(a) Find the first seven decimal places of $\sqrt{2}$ and $\sqrt{3}$ exactly.

(b) Put $l = \sqrt{3}$ and $b = \sqrt{2}$. Let $l_i$ denote the number formed by the first $i$ decimal places of $l$; $b_i$ is defined similarly. Find $l_1, l_2, \ldots, l_7$, and $b_1, b_2, \ldots, b_7$.

(c) Use the fact (proved in the text) that $a_i = l_i \times b_i$ to calculate $a_1, a_2, \ldots, a_7$ exactly.

(ii) In the text, A'B'C'D' is a "16/13 by 2" rectangle, and $b_0, b_1, b_2, b_3, \ldots$ is the corresponding sequence of inner approximations which pinpoints the number $\beta = \text{area A'B'C'D'}$. (N.B., these $b_i$'s have nothing at all to do with the $b_i$'s in part (i).) Use the method of part (i) to find $b_1, b_2, \ldots, b_7$ exactly.

(iii) Find the first six decimal places of $\alpha/\beta$ exactly.

10. When we applied the method of inner and outer approximations to general shapes in 2-dimensions we obtained

an increasing sequence $a_0$, $a_1$, $a_2$, ... of inner approximations, which pinpoints the real number $\alpha$, and
a decreasing sequence $A_0$, $A_1$, $A_2$, ... of outer approximations, which pinpoints the real number $A$.

We showed that

$$a_0 \leq a_1 \leq a_2 \leq \ldots \leq \alpha \leq \ldots \leq A_2 \leq A_1 \leq A_0$$

and

$$a_0 \leq a_1 \leq a_2 \leq \ldots \leq A \leq \ldots \leq A_2 \leq A_1 \leq A_0.$$

Show that, if

$$A < \alpha,$$

then there exists an $N$ such that

$$A < a_N.$$

Hence conclude that

$$\alpha \leq A.$$

11. (i) Dissect the unit interval $[0, 1]$ (Figure 92) into its separate points: $\{x: 0 \leq x \leq 1\}$. Then put these separate points back together again in the following way: place each point "$x$" at the position "$2x$" on the number line

Figure 92

(Figure 93); thus the point "0" is replaced at the point "0", and the point "1/3" is replaced at the point "2/3". Show that when the separate points of $[0, 1]$ are reassembled in this way, they fill up the complete interval $[0, 2]$. *Hence this dissection and rearrangement does not preserve lengths.*

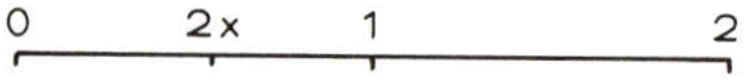

Figure 93

(ii) The apparently paradoxical result in part (i) can be explained by observing that we have dissected the interval $[0, 1]$ *into too many pieces*. In Exercise 7 we introduced the radically new idea that we might dissect a shape Y into an *endless sequence*

$$Y_1, \quad Y_2, \quad Y_3, \quad Y_4, \ldots$$

of separate pieces and still find that the length (or area) of Y was equal to the *endless* (!) sum of the lengths (or areas) of the separate pieces:

$$\text{length(Y)} = \sum_{i=1}^{\infty} \text{length}(Y_i). \tag{*}$$

Now if Y = [0, 1], then

$$\text{length}(Y) = 1.$$

When we dissect [0, 1] into its *separate points* "*x*," the length of each point is clearly *zero*!

But an endless sum of zeros still gives zero (remember that the value we assign to an endless sum is precisely that number which is pinpointed by the corresponding endless sequence of finite sums, and these finite sums are all zero).

Does it not then follow from the equation marked (*) that

$$l = \text{length}[0, 1] = \sum_{0 \leqslant x \leqslant 1} \text{length}(\text{"}x\text{"}) = 0?$$

Fortunately not! For *there are in fact far too many points in the interval [0, 1] for us to be able to include them all in a single endless sequence.*

Thus the sum

$$\sum_{0 \leqslant x \leqslant 1} \text{length}(\text{"}x\text{"})$$

is not a proper endless sum, and so has (as yet) no meaning.

12. Let X be an "*l* by *b*" rectangle. If X is reduced to a standard rectangle X' by the dissection method described in the text, show that X' is a "1 by *l* · *b*" rectangle. [Hint:

   (i) Show that the bisection step does not change the value *l* · *b* of the product of the lengths of the two sides. Thus we may assume that $l \geq 1$ and $b \leq 1$.

   (ii) If X is the rectangle ABCD, and AF has length 1 (Figure 94), find similar triangles which allow you to conclude[15] that the length of AG is equal to $l \cdot b$.]

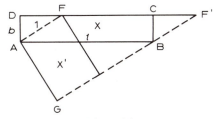

Figure 94

13. (i) It is clear that a *regular* pentagon can be dissected by a diagonal into a triangle and a quadrilateral (Figure 95). In this instance life is far too kind: we cannot possibly go wrong, since *every* diagonal cuts the pentagon into a triangle and a quadrilateral. But when working with other pentagons we will usually

---

[15] You should no longer be surprised to learn that in order to prove the familiar properties of similar triangles, one needs a geometrical version of the fundamental property of real numbers!

Figure 95

have to choose more carefully if we want a diagonal which will cut off a triangle
and leave a quadrilateral (Figure 96). Try to produce some kind of proof that

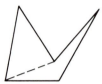  

Figure 96

every pentagon can be dissected by a diagonal into a triangle and a quadrilateral provided we choose the diagonal carefully.

(ii) It may seem intuitively obvious that every $(n + 1)$-gon $A_0 A_1 A_2 \ldots A_n$ can be
cut into

$$\text{a 3-gon } A_i A_{i+1} A_{i+2} \text{ and an } n\text{-gon } A_{i+2} \ldots A_n A_0 \ldots A_i.$$

But it is not at all clear which three successive vertices

$$A_i, \quad A_{i+1}, \quad A_{i+2}$$

actually form a protruding triangle which can be cut off without interfering with
the rest of the $(n + 1)$-gon: for example, when $n = 8$ we may be confronted by
the $(8 + 1)$-gon in Figure 97, in which, of the nine possible triples $A_i A_{i+1} A_{i+2}$,

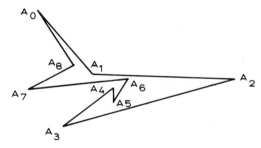

Figure 97

only two

$$A_8 A_0 A_1 \quad \text{and} \quad A_3 A_4 A_5$$

can actually be cut off ($A_4 A_5 A_6$ does not count, since that protrudes inwards).

It is conceivable that, for larger values of $n$, there may exist very complicated
$(n + 1)$-gons which have *no suitable protruding triangles at all*. Prove that this
cannot happen, and that in every $(n + 1)$-gon (for every value of $n \geq 3$) there
exist at least two suitable protruding triangles.

(iii)  Use part (ii) to prove (by induction on $n$) that every polygon $A_0 A_1 A_2 \ldots A_n$ can be reduced to a standard rectangle.

14. What is the largest number of ".1 by .1" squares which can be fitted (without overlapping) inside a circle of radius 1?

15. (i)  Try to prove that *every polygon in the plane has an "inside" and an "outside"* (see Figure 98). [Hint: A polygon *in space* certainly has no "*inside*." Thus it

Figure 98

seems that a proof must somehow capture the 2-dimensional idea that to get from the "*inside*" to the "*outside*" in the plane one has to cross the polygon itself ... .]

(ii)  There seems to be no elementary proof of the fact that *every "closed curve" in the plane has an "inside" and an "outside."* At what point (or points) does your proof break down for general "closed curves"?

# Comparing Volumes

In Chapter III.3 we discovered that comparing shapes in 2-dimensions was noticeably more complicated than comparing plain line segments. The proverbial optimist might of course declare that we should have expected *1-dimension* to be rather special, and that, now we know (more or less) how to make the jump from 1- to 2-dimensions, we shall probably find that 3-, 4- and higher dimensions are really no more difficult than 2-dimensions. The pessimist, on the other hand, might point out that, since 2-dimensions gave rise to so many unexpected difficulties, we must surely expect 3-, 4- and higher dimensions to become steadily more complicated.

In the event each of these views turns out to be partially justified. Our two approaches to measuring shapes in 2-dimensions can be adapted so as to allow us to measure shapes in 3-dimensions; but the extra degree of freedom in 3-dimensions does give rise to some unexpected difficulties which force us to make fairly major changes, even for very simple shapes such as pyramids and tetrahedra. In this section we make no attempt to discuss the 3-dimensional problem in detail; instead, we simply indicate how the ideas of the previous section can be extended to 3-dimensions, and hint at some of the unexpected complications which accompany the step from 2- to 3-dimensions.[1]

We begin by investigating how each of our two approaches actually gets started in 3-dimensions. In 2-dimensions both the method of inner and outer approximations and the dissection approach *started* by considering rectangles, the fundamental importance of which is reflected in the simplicity of the familiar

$$length \times breadth$$

area formula.[2] It does not require much geometric imagination to guess the most promising starting point for our study of volume in 3-dimensions:

---

[1] We have incidentally seized the opportunity to exercise the reader's 3-dimensional *imagining*, by exploring some of the interesting, but elementary, problems related to our main task.

[2] This formula was essentially proved in the previous chapter by each method *separately* (the proof via the dissection approach follows from Exercise 12 and Exercise 1 of Chapter III.3).

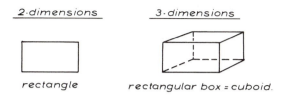

Figure 99

namely, the *rectangular box* usually called a *cuboid* (Figure 99). In 2-dimensions the method of inner and outer approximations measures

first in terms of "1 by 1"squares

then in terms of ".1 by .1" squares

and so on: hence we would expect the analogous method in 3-dimensions to measure

first in terms of "1 by 1 by 1" cubes

then in terms of ".1 by .1 by .1" cubes

and so on.

So far then, so good. But how should we formulate a 3-dimensional analogue of the dissection method? In 2-dimensions the dissection method sought to reduce each given shape $X$ to a "1 by *something*" *standard rectangle*. What then should we take as the 3-dimensional analogue of a standard rectangle? Recall that the dissection approach usually reduced the initial shape $X$ to two (or more) standard rectangles—say Y' ("1 by $y$") and Z' ("1 by $z$") (Figure 100), which we then put together to make a single

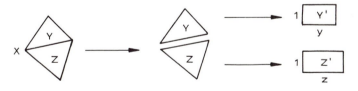

Figure 100

"1 by $(y + z)$" standard rectangle. This stresses the fact that it is the *uniform cross-section* which is most important, and suggests that the term *standard cuboid* should mean precisely

a "*1 by 1 by something*" *cuboid.*

Thus, given a 3-dimensional shape X, the dissection approach requires us to find a way of cutting it up and reassembling the pieces to make a standard cuboid X'.

Now that we think we understand the basic strategy behind each approach in 3-dimensions, we must next investigate the problem of comparing two arbitrary cuboids X (=ABCDEFGH) and Y (=A'B'C'D'E'F'G'H') (Figure 101).

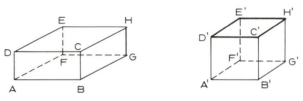

Figure 101

Let us start by considering the easiest case before looking at the general problem.

If AB, A'B' have a common measure WX,
and AD, A'D' have a common measure WZ,
and AF, A'F' have a common measure WT,

*then* we can construct a "mini-cuboid" M (=WXYZSTUV), which is a common measure for the two original cuboids X and Y, and which fits precisely

$$\frac{AB}{WX} \cdot \frac{AD}{WZ} \cdot \frac{AF}{WT} \text{ times into X}$$

and

$$\frac{A'B'}{WX} \cdot \frac{A'D'}{WZ} \cdot \frac{A'F'}{WT} \text{ times into Y.}$$

Therefore

$$\frac{X}{Y} = \frac{\text{volume X}}{\text{volume Y}} = \frac{AB}{A'B'} \cdot \frac{AD}{A'D'} \cdot \frac{AF}{A'F'}.$$

In particular, if Y happens to be a unit cube, then we have proved the familiar

*length × breadth × height*

formula for the volume of a cuboid X *which has all edges of rational length.*

Observe that we have not yet used *the full force* of either of our two general methods.[3] However, if one (or more) of the pairs of edges (AB, A'B'), (AD, A'D'), (AF, A'F') has no common measure, then our first alternative is to resort to the full-blooded method of inner and outer approximations to measure each of X and Y.

---

[3] We have used both methods to some extent. On the one hand our argument may be interpreted as *approximating* each cuboid (X and Y) *both inside and out* with copies of the mini-cuboid WXYZSTUV in place of the usual "1 by 1 by 1" cubes; on the other hand the argument depends on *dissecting* both X and Y into copies of the mini-cuboid WXYZSTUV.

We begin by measuring both X and Y in terms of "1 by 1 by 1" cubes: this gives us crude approximations ($a_0$, $A_0$ for X, and $b_0$, $B_0$ for Y)

$$a_0 \leq \text{volume X} \leq A_0, \qquad b_0 \leq \text{volume Y} \leq B_0.$$

Next we measure both X and Y in terms of ".1 by .1 by .1" cubes, which gives us better approximations ($a_1$, $A_1$ for X and $b_1$, $B_1$ for Y)

$$(a_0 \leq )a_1 \leq \text{volume X} \leq A_1(\leq A_0),$$
$$(b_0 \leq )b_1 \leq \text{volume Y} \leq B_1(\leq B_0);$$

and so it goes on. The terms of the endless increasing sequence

$$a_0 \leq a_1 \leq a_2 \leq a_3 \leq \ldots$$

are all less than or equal to $A_0$, so by the fundamental property of real numbers this sequence pinpoints a real number $\alpha$. Similarly the terms of the endless decreasing sequence

$$A_0 \geq A_1 \geq A_2 \geq A_3 \geq \ldots$$

are all greater than or equal to $a_0$, so this sequence pinpoints a real number $A$. *Because X is a cuboid* one can prove (Exercise 2) that the two real numbers $\alpha$ and $A$ which are pinpointed in this way are always equal—a fact which can be used to show what we quietly assumed above: namely that the cuboid X actually has a "*volume.*" (Strictly speaking, we should first define the $a_i$'s and $A_i$'s without mentioning the "*volume*" of X, then show that the numbers $\alpha$ and $A$ pinpointed by these two sequences are equal, and finally *define*

$$\text{``volume'' of X} = \alpha = A.)$$

This method not only provides a way of obtaining a *numerical* value for the volume of each particular cuboid X, but also allows us to prove the familiar volume formula

$$\textit{length} \times \textit{breadth} \times \textit{height}$$

for *every* cuboid X (Exercise 3).

Our second alternative is to find a way of dissecting an arbitrary "$x$ by $y$ by $z$" cuboid in such a way that the pieces can be reassembled to make a standard "1 by 1 by ?" cuboid; this requires a little thought, but it is an excellent elementary exercise in 3-dimensional thinking (see Exercise 4). We can then reduce each of the two cuboids X and Y, which we originally wanted to compare, to standard cuboids

$$X'(\text{``1 by 1 by } \alpha\text{''}) \quad \text{and} \quad Y'(\text{``1 by 1 by } \beta\text{''}).$$

The problem of comparing X' and Y' is now essentially a 1-dimensional problem, so we can use the method of Chapter II.13 (see Exercise 1 of Chapter III.3) to show that

$$\frac{X}{Y} = \frac{X'}{Y'} = \frac{\alpha}{\beta}.$$

In particular, if Y' happens to be a "1 by 1 by 1" cube, this shows that the volume of the "1 by 1 by $\alpha$" standard cuboid X' is precisely equal to its length $\alpha$; if we combine this remark with the fact (Exercise 5) that our dissection procedure turns an

"$x$ by $y$ by $z$" cuboid

into a

"1 by 1 by $xyz$" standard cuboid

then we have a proof of the

*length* × *breadth* × *height*

formula for the volume of an "$x$ by $y$ by $z$" cuboid.

We may therefore assume that we know how to measure the volume of any cuboid by each of our two methods. But where do we go from here?

In 2-dimensions we simply stated—without proof— that the method of inner and outer approximations could, in principle, be used to measure the area of *any polygon*. By this we meant no more, and no less, than that we could give a theoretical proof that

*if X is a polygon,*
*then the real number $\alpha$ pinpointed by the increasing sequence*

$$a_0, \quad a_1, \quad a_2, \quad a_3, \ldots$$

of inner approximations to X,
   *and the real number A pinpointed by the decreasing sequence*

$$A_0, \quad A_1, \quad A_2, \quad A_3, \ldots$$

of outer approximations to X
   *are necessarily equal*, and

$$\alpha = A = \text{area } X.$$

But we also stated fairly clearly that, when it comes to actually calculating areas of polygons, this approach is rather hard to use as it stands.[4] Exactly the same can be said of the 3-dimensional version. However we should

---

[4] However, if we use a little imagination and break away from our preoccupation with *squares* as the means of approximation, then this approach and its refinements become reasonably effective calculating devices (Exercise 6).

stress that, *in spite of its apparent weaknesses, this approach and its subsequent refinements lead eventually to the arithmetical descriptions of both the Riemann and Lebesgue integrals, on which modern analysis is based.*

But though the method of inner and outer approximations and its refinements are important, they are certainly not all-important: there are other, more elementary, ideas which have contributed to the development of the naive concept of the integral. One of these ideas stands out as being particularly worthy of attention in any introductory analysis of infinite processes. We shall therefore begin by looking at the unexpected difficulties encountered by the dissection approach in 3-dimensions, and at the reasons why infinite processes are needed to handle even very simple shapes—such as pyramids and tetrahedra. Finally we shall consider the naive, but effective *"method of slices,"* from which the naive version of the integral emerged in the course of the sixteenth and seventeenth centuries.

In 2-dimensions once we had discovered how to reduce an arbitrary rectangle to a standard rectangle, we turned our attention to other 2-dimensional shapes *formed wholly by straight lines*—first

*parallelograms*

then

*triangles*

and finally the most general 2-dimensional shapes formed wholly by straight lines, namely

*general polygons.*

But what is the 3-dimensional analogue of this

rectangle–parallelogram–triangle–general polygon

sequence which worked so well in 2-dimensions?

The most general 3-dimensional shapes *formed wholly by straight lines* would seem to be

*general polyhedra*[5] (Figure 102).

---

[5] Though we conveniently used the expression "3-dimensional shape *formed wholly by straight lines*" as in 2-dimensions, we should strictly have bumped the words in italics up one dimension also:

the simplest 0-dimensional shape is a *point*;
a 1-dimensional shape "bounded by" *points* is a *line segment*;
a 2-dimensional shape "bounded by" *line segments* is a *polygon*;
a 3-dimensional shape "bounded by" *polygons* is a *polyhedron*;
a 4-dimensional shape .... .

Even this falls short of the whole truth—just how far short can be seen from I. Lakatos' very readable book: *Proofs and Refutations*, (CUP, 1977).

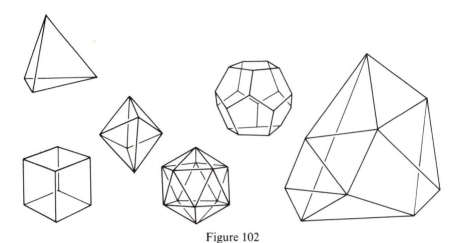

Figure 102

So, we know how to reduce *any cuboid* to a standard cuboid by dissection, and we would like to be able to reduce *any polyhedron* to a standard cuboid by dissection.

Look back at how we made the analogous jump

from *rectangles*        to *general polygons*

in 2-dimensions. There our first move was to show that any *parallelogram* X could be reduced to a rectangle X′ (Figure 103). We were then in a position to exploit the fact that we already knew how to reduce the rectangle X′ to a standard rectangle.

So what is the 3-dimensional analogue of all this? One might at first assume that since a parallelogram arises by "deforming," or "shearing," a rectangle (Figure 104), so its 3-dimensional analogue should be the shape obtained by "deforming," or "shearing," a cuboid in the same way (Figure 105). This is especially encouraging since it is easy to see how such a shape can be reduced to a cuboid (Figure 106).

However our original deformation of the cuboid was unnecessarily restricted. We fixed the base ABGF and sheared parallel to the edge AB; we thus deformed the rectangle ABCD into a parallelogram ABC′D′, and the

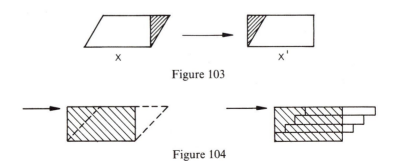

Figure 103

Figure 104

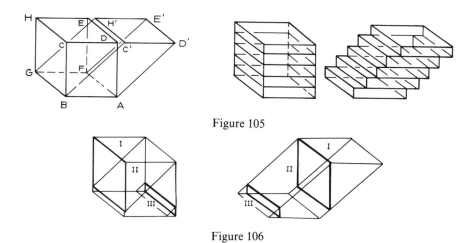

Figure 105

Figure 106

rectangle FGHE into a parallelogram FGH′E′—but the faces ABGF and DCHE remained exactly the same shape as before. And though the faces ADEF and BCHG certainly changed shape, their images AD′E′F and BC′H′G *are still rectangular!*

If however we interpret the 2-dimensional step

from *rectangles*

to *parallelograms*

as a step

from arbitrary rectangular coordinates

to arbitrary oblique coordinates (Figure 107),

then it is clear that the corresponding step in 3-dimensions should be

from arbitrary rectangular coordinates

to arbitrary oblique coordinates (Figure 108);

in other words

from *cuboids*

to *parallelepipeds.*

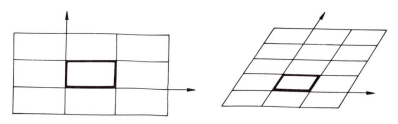

Figure 107

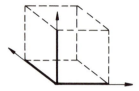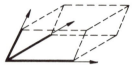

Figure 108

The only restriction on the six faces of a parallelepiped is that they should be *parallel in pairs*, from which it follows that all six faces are parallelograms and that parallel faces are congruent (Exercise 7).

Fortunately, each parallepiped X can be reduced to a cuboid X'—though it will probably not be immediately obvious how, and you should try to find the simplest way before turning to Exercise 8. The cuboid X' can then be reduced to a standard cuboid X''. Hence each parallelepiped can be reduced to a standard cuboid.

Recall that our aim is to reduce *any polyhedron* to a standard cuboid; our progress so far can be summarised thus:

*cuboid–parallelepiped–?–?– ... –general polyhedron.*

But before we begin to feel too pleased with ourselves we should remember that, in 2-dimensions, it was the next step (from *parallelograms* to *triangles*) which turned out to be the really crucial step! The method we used to reduce a general $(n + 1)$-gon to a standard rectangle started by *cutting off a vertex* from the $(n + 1)$-gon (Figure 109), thereby reducing it to a triangle and an *n*-gon. But we did not stop there. The triangle can be

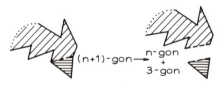

Figure 109

easily reduced to a rectangle, and so to a standard rectangle. But the *n*-gon has to undergo exactly the same kind of surgery as the $(n + 1)$-gon

$$(n + 1)\text{-gon} \longrightarrow n\text{-gon} \longrightarrow (n - 1)\text{-gon} \longrightarrow \ldots$$
$$+ \qquad\qquad + $$
$$\text{triangle} \qquad \text{triangle} \qquad \ldots$$

Thus in 2-dimensions our method worked only because we managed to complete *both* of the following steps:

*Step 1 :* Dissect each $(n + 1)$-gon into $(n - 1)$ triangles.
*Step 2 :* Reduce each triangle to a standard rectangle.

If we try to imagine the analogous 3-dimensional process of "cutting off a vertex from a polyhedron with $n + 1$ vertices" (Figure 110), then it looks as if we should expect the polyhedron with $n + 1$ vertices to be dissected into a polyhedron with $n$ vertices, together with some kind of *pyramid*.

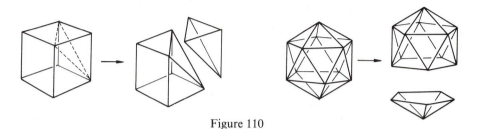

Figure 110

Thus our first guess at the correct analogue of

*a triangle* in 2-dimensions

would seem to be

*a general pyramid* in 3-dimensions.

If this were the case, then our next step (analogous to *Step 2* above) should be to find a way of reducing each such pyramid to a standard cuboid. The reasonableness of this first guess is strongly supported by the similarity between the familiar formula

$$\frac{1}{2} \text{ (length of base)} \times \text{height}$$

for the area of a *triangle* in *2*-dimensions, and the formula

$$\frac{1}{3} \text{ (area of base)} \times \text{height}$$

for the volume of a *pyramid* in *3*-dimensions.

This first guess is essentially correct, but in most mathematics it is more natural to take, not general pyramids, but rather *tetrahedra* (Figure 111);

Figure 111

(that is, pyramids with a triangular base) as being the true 3-dimensional analogue of triangles in 2-dimensions. This makes very little difference in practice, because any pyramid with an $(n + 1)$-gon as base can be dissected into $n - 1$ tetrahedra (Exercise 9). Thus if only we can find a way of reducing each tetrahedron to a standard cuboid, then we shall be able to reduce every pyramid to a standard cuboid; first dissect the pyramid into $n - 1$ tetrahedra $T_1, T_2, \ldots, T_{n-1}$; then reduce each tetrahedron $T_i$ to a standard "1 by 1 by $\alpha_i$," cuboid (Figure 112), and finally put all these standard cuboids end to end—thereby reducing the original pyramid to a "1 by 1 by $(\alpha_1 + \alpha_2 + \ldots + \alpha_{n-1})$" standard cuboid.

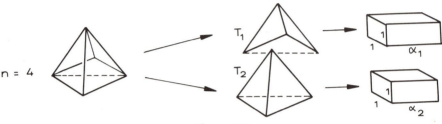

Figure 112

Unfortunately there are two things wrong with our naive approach. Firstly, we made a fuss in 2-dimensions as to whether it was always possible to "cut off a vertex from an $(n + 1)$-gon" in such a way as to obtain a triangle and an $n$-gon. The problem was not so much that we believed the assertion to be false, but rather that it is not at all easy to find a clearly convincing proof (see Exercise 10 (i)). In contrast, our naive assertion, that in 3-dimensions one can always "cut off a vertex of a polyhedron with $n + 1$ vertices" in such a way as to obtain a pyramid and a polyhedron with $n$ vertices, is easily seen to be *false* (Exercise 10 (ii))! Fortunately one can show (Exercise 11) that *every convex*[6] *polyhedron with $n + 1$ vertices can be dissected into tetrahedra*, but the proof is a bit tricky and there is no nice formula for the number of tetrahedral pieces one gets. Nevertheless, let us be thankful for small mercies: the fact that such a polyhedron can be dissected into tetrahedra at least guarantees that, *once we discover a way of reducing any tetrahedron to a standard cuboid*, we shall automatically have a way of reducing any (convex!) polyhedron to a standard cuboid.

The problem of reducing any pyramid to a standard cuboid remained unsolved until 1900, when David Hilbert proposed it as the third of the twenty three problems which he presented at the International Congress of Mathematicians in Paris. In our language, Hilbert's third problem was essentially this:

Problem III:    *Can a tetrahedron of unit volume always be
                 reduced to a standard "1 by 1 by 1" cuboid?*

---

[6] A polyhedron is convex *if it has no "dents"*: that is, if whenever A, B are two points on the surface of the polyhedron, then all the points which lie on the line segment AB between A and B lie *inside* the polyhedron.

The twenty-three problems, and the attempts of many many mathematicians to solve them, have had a considerable influence on the development of mathematics in the twentieth century.[7] Problem III was certainly the easiest of Hilbert's problems, and was solved in the same year (1900) by Max Dehn (1875—1952), who proved that *it is impossible to dissect a regular tetrahedron into a* finite *number of pieces which can be reassembled to form a cube.*

One slightly surprising consequence of this result is that although the formula "$\frac{1}{2}$(length of base) × height" for the area of a triangle is very easy to prove, and although the formulae

$$\text{"}\tfrac{1}{2}\text{(length of base)} \times \text{height,"}$$

and

$$\text{"}\tfrac{1}{3}\text{(area of base)} \times \text{height"}$$

are tantalisingly similar, there can be no elementary proof of the formula for the volume of a tetrahedron! Every proof must involve infinite processes in some possibly disguised form.

This raises an intriguing historical question. Interest in the volumes of pyramids and cones goes back long before the invention of even the crudest infinite processes. How then did the mathematicians of the time set about calculating such volumes?

Just as the formula for the area of a triangle is equivalent to the formula for the area of a trapezium (Figure 113), so also the formula for the volume

$$\frac{1}{2} b \, (h_1 + h_2) \quad = \quad \text{area } X \; + \; \frac{1}{2} a . h_2$$

$$\therefore \; \text{area } X \quad = \quad \frac{1}{2} (bh_1 + bh_2 - ah_2)$$

$$= \quad \frac{1}{2} (a+b).h_1 \, , \; \text{since} \; \frac{h_1 + h_2}{b} = \frac{h_2}{a}$$

Figure 113

[7] The resolution in recent years of some of the hardest and most important problems has led to the publication of two notable books surveying progress towards the solution of all twenty three. A Russian survey appeared in 1969, and was later translated into German (*Die Hilbertschen Probleme*, Akademische Verlagsgesellschaft, Leipzig, 1979). A two volume survey in English (*Mathematical Developments Arising from Hilbert's Problems*, American Mathematical Society, 1976) includes an English translation of Hilbert's original talk: most of the reports on individual problems are highly technical, but it is still worth browsing through them to get a feeling for the scope both of Hilbert's vision and of modern mathematics.

of a pyramid is equivalent to the formula for the volume of a *truncated* pyramid (Figure 114). So if one does not know how to calculate the volume

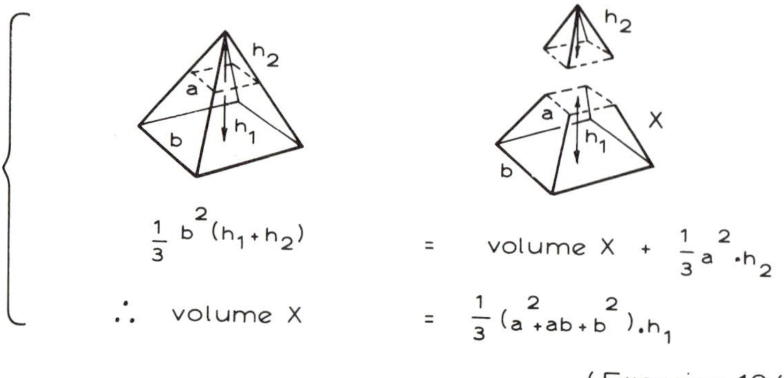

$$\frac{1}{3}b^2(h_1+h_2)$$

$$= \quad \text{volume X} \quad + \quad \frac{1}{3}a^2 \cdot h_2$$

$$\therefore \quad \text{volume X} \quad = \quad \frac{1}{3}(a^2+ab+b^2) \cdot h_1$$

(Exercise 12 (i))

Figure 114

of a square pyramid, then one cannot calculate the volume of a truncated square pyramid either. One may of course resort to actual physical measurements to find approximate "rule of thumb," or one may try to guess the 3-dimensional result by looking at the 2-dimensional result for the area of a trapezium. In some Babylonian texts (around 2000 BC) we find a rule for the volume of what they call a "basket," which historians have interpreted as being a truncated cone or a truncated square pyramid. This rule is obtained in the spirit of the formula $\frac{1}{2}(a + b) \cdot h$ for the area of a trapezium by averaging the areas of the two parallel faces:

volume of "basket" = (average of parallel faces) × height.

For the volume of a truncated square pyramid, whose parallel faces are separated by a distance $h$ and have edge lengths $a$ and $b$ respectively (Figure 115), this gives rise to a rule which we would express algebraically in the form

$$\tfrac{1}{2}(a^2 + b^2) \times h.$$

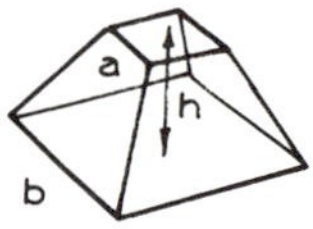

Figure 115

An alternative guess in the same spirit might try to generalize the fact that the formula $\frac{1}{2}(a + b) \times h$ for the area of a trapezium can also be interpreted as

*(horizontal cross-section half way up)* × *height*;

this would give rise to a rule for the volume of a truncated square pyramid which we would express algebraically in the form

$$[\tfrac{1}{2}(a + b)]^2 \times h.^8$$

Another old-Babylonian text contains a calculation based on a rule which begins rather like this, but which adds a 'correction term': the interpretation of this correction term is obscure, but Neugebauer has suggested that it represents an application of a (perfectly correct) rule which we would express algebraically in the form

$$\left( [\tfrac{1}{2}(a + b)]^2 + \frac{1}{3} \left[ \frac{a - b}{2} \right]^2 \right) \times h.$$

Whether Neugebauer's interpretation is correct, or how such a correction term might have been discovered, we simply do not know.[9] The Moscow papyrus (around 1850 BC) shows clearly that, about the same time, Egyptian mathematicians were also in possession of a correct rule $\frac{1}{3}(a^2 + ab + b^2) \cdot h$ for the volume of a truncated square pyramid.

But to return to our task. We now know that in order to extend our discussion of volume to cover tetrahedra and pyramids, we have no choice but to introduce infinite processes: we must therefore decide *the form in which infinite processes should be incorporated into our approach to volume.*

Perhaps the most straightforward beginning would be to say exactly what we mean by a dissection of the shape X into an *endless sequence of pieces*

$$X_1, \quad X_2, \quad X_3, \quad X_4, \ldots .$$

There is no real difficulty here, and the approach is very important since it leads to the notion of Lebesgue measure (which generalises our idea of volume) and the Lebesgue integral (which generalises our idea of integration). It would however be premature to take up such an approach at this stage. Instead, for the remainder of this section, we shall throw caution and precision to the winds to consider

*the method of slices*

which was in many ways responsible for the original emergence of integration.

---

[8] In *A History of Mathematics* (Wiley, 1968), C. B. Boyer states that this rule was actually used in some Babylonian texts, but he gives no reference and I have been unable to substantiate the claim.

[9] For a refreshingly clear discussion of the evidence for this, and many other aspects of ancient mathematics, see B. L. van der Waerden's beautiful book, *Science Awakening* (published by Noordhoff, 4th edition 1975).

As the name suggests, the method of slices approaches each question about the volume of a figure in 3-dimensions, or about the area of a figure in 2-dimensions, by considering certain carefully chosen cross-sections or "slices." This has an intuitive appeal which is undeniable, but there is no way in which the method, as it stands, can be squared with either the two approaches to volume described in this chapter, or the two approaches to area described in the previous chapter. Both the method of inner and outer approximations and the dissection approach are based on the idea that

the *volume* of any figure in 3-dimensions

is determined by

the *volumes* of simpler pieces into which it can be dissected

(such as "$1/10^n$ by $1/10^n$ by $1/10^n$" cubes together with a remainder). Similarly

the *area* of any figure in 2-dimensions

is determined by

the *areas* of simpler pieces into which it can be dissected.

In contrast the method of slices appears to be thoroughly confused, in that it suggests that

the *volume* of a figure in 3-dimensions

is somehow determined by

the *areas* of certain cross-sections or slices,

and that

the *area* of a figure in 2-dimensions

is somehow determined by

the *lengths* of certain cross-sections or slices.

This may no longer shock those who have become used to calculating the *area* under the curve $y = f(x)$ by integrating the *length* $f(x)$ of the cross-section perpendicular to the $x$-axis (Figure 116). But it should! There is, on the face of it, no earthly reason why the area of the shaded semicircle in Figure 117 should be equal to the shaded area between the cycloid and Roberval's auxiliary curve (see Exercise 14); neither is there any reason to expect a cone with elliptical base, having major and minor axes of length $a$ and $b$ respectively, to have the same volume as a pyramid with a "$\pi a$ by $b$"

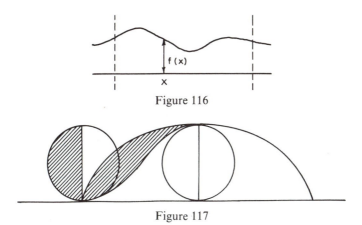

Figure 116

Figure 117

rectangular base and the same height (Figure 118). But the method of slices, when carefully used, is so effective, that even those who recognise its lack of pedigree must suspect that there is some underlying theoretical reason *why* it works.

Figure 118

The method was presumably used in some form long before the time of Archimedes (287?–212 B.C.): but it was his supreme geometrical insight which convinced mathematicians of the sixteenth and seventeenth centuries that the method might be developed into a universal "calculus." [10] Most of Archimedes' applications have an element of surprise and elegance which still takes one's breath away. We shall look at the ingenious way he dis-

---

[10] The influence exerted by Archimedes in promoting the method of slices was largely indirect. It was widely known that he had used some such method *to discover* many of his most impressive results, but no details were known. This was largely because Archimedes realised that, however suggestive the method might be, it could never constitute a proof: he therefore *re-proved* all his results in a "strictly correct" way before publishing them. For example, he gave a strictly correct proof relating the volumes of sphere, cone, and cylinder, in which the result itself comes at the end of forty odd pages, after some thirty-four preliminary Propositions. Details of the method by which Archimedes discovered his results only became known in 1906 when J. L. Heiberg discovered, in what is now Istanbul, a copy of a letter from Archimedes to Eratosthenes. But though precise details were previously lacking, Archimedes' whole approach shines through many of his other works which were widely available in the late sixteenth and early seventeenth centuries. Thus, for example, Roberval (1602–1675) declares that it was by studying the "divine Archimedes" that he discovered how to use the method of slices to find the area under the cycloid (Exercise 14).

covered the correct formula for the volume of a sphere; but to appreciate it fully we need first a short paragraph of background.

The fact that the area of a circle is *proportional to the square of the radius* was not only known, but had been rigorously proved in Euclid's *Elements* (Book XII, Proposition 2). No precise value for the "constant of proportionality" was known, so we shall simply call it $\pi$.[11] In the same Book XII Euclid essentially proved the rules for the volume of a pyramid (Propositions 3–7), for the volume of a cone (Proposition 10), and for the volume of a circular cylinder (Proposition 11 and its proof).[12] At the end of Book XII Euclid proves that the volume of a sphere is proportional to the cube of the radius, *but he says nothing at all about the mysterious constant of proportionality*: in particular he does not suggest any connection at all between

$$\text{the constant} \quad \frac{\text{volume of sphere}}{\text{cube of radius}}$$

and

$$\text{the constant} \quad \frac{\text{area of circle}}{\text{square of radius}}.$$

Archimedes' discovery was simply that

*a hemisphere of radius r* and *a right circular cone with base-radius and height r*

taken together have the same volume as

*a cylinder of radius r and height r.*

So since the volume of the cone $\frac{1}{3}\pi r^3$ and the volume of the cylinder $\pi r^3$ were already known, Archimedes concluded that the volume of the hemisphere must be $\frac{2}{3}\pi r^3$ and thus established a totally unexpected connection between the two constants

$$\frac{\text{area of circle}}{\text{square of radius}} = \pi \quad \text{and} \quad \frac{\text{volume of sphere}}{\text{cube of radius}} = \frac{4}{3} \cdot \pi.$$

---

[11] The definition of $\pi$ as "*area of circle/square of radius*" allowed them to calculate the value reasonably accurately: thus, for example, Archimedes used the areas of inscribed and circumscribed polygons to show that

$$3\frac{10}{71} < \pi < 3\frac{1}{7}.$$

[12] This is not strictly true since Euclid was not interested in "rules" or formulae. But what he does prove in the quoted Propositions can be formulated in this way for our purposes.

(In Exercise 15 you are invited to continue this sequence into 4-dimensions and beyond!)

And how did Archimedes convince himself that the total volume of cone and hemisphere taken together was equal to that of the cylinder? He arranged the cone and hemisphere as shown in Figure 119, and then simply observed that, at each level $x$ above the base,

*the "amount"* ( = *cross-sectional slice*) *of cylinder* ( = $\pi r^2$)

exactly balances

*the "amount" of hemisphere* ( = $\pi(r^2 - x^2)$) and *the "amount" of cone*
( = $\pi x^2$).

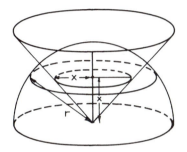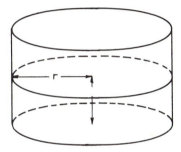

Figure 119

We shall end this chapter by showing how the naive method of slices could be used to discover the $\frac{1}{3}$(*area of base*) × *height* formula for the volume of a tetrahedron.

Look again at the picture of a cuboid which has been deformed by a simple shear (Figure 120). The second version of this picture suggests very strongly the idea that such a deformation simply slides "successive slices" relative to one another—much as one might do with a pack of cards—and that since the volume of each slice remains unchanged, the total volume must also remain unaltered. The weakness of the argument lies in the fact that, to talk about the "volume of each slice" requires that each slice should have positive thickness; but in that case the stack of slices is never quite the right shape. In the special case where we are deforming a cuboid in this way the difficulty can be resolved: simply cut off the offending jagged pro-

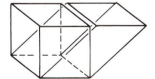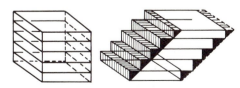

Figure 120

trusions (shaded) from the left of each slice and use them to fill the identical jagged holes on the right of each slice. In the case of the tetrahedron one simply has to accept that there is no such easy way out: we use the idea for its suggestive power only.

Take any tetrahedron WXYZ and choose one face as base, say WXY of area A. Let the corresponding height of Z above WXY be $a$ (Figure 121).

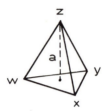

Figure 121

According to the method of slices, we may deform the tetrahedron *without changing its volume* provided we keep the base fixed and move the apex in the plane parallel to the base—that is, in such a way as to preserve the height (Exercise 16(i)). This means, in particular that any two tetrahedra with congruent bases, and corresponding heights equal, always have equal volumes. We shall use this fact later on. Meantime move the apex Z in the plane parallel to WXY to the point Z' vertically above one of the base vertices—say X (Figure 122). This produces two right angled triangles

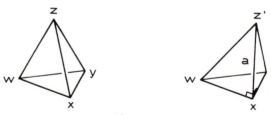

Figure 122

WXZ' and YXZ' sharing a common edge XZ' of length $a$. Next tilt the tetrahedron onto the new base XYZ' (Figure 123). The original base WXY

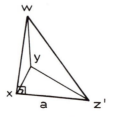

Figure 123

is now a vertical face with base XY and height equal to the height of the tetrahedron: hence if W is moved parallel to XY to the point W' vertically over X (Figure 124), the area of the new face W'XY is equal to the area A of the original base WXY, the volume of the new tetrahedron W'XYZ' is equal to the volume of the original tetrahedron WXYZ, and the height XZ' = a

Figure 124

corresponding to the base W'XY of the new tetrahedron is equal to the height a of the original tetrahedron (relative to the face WXY as base). Moreover, the three angles Z'XW', W'XY, YXZ' of the new tetrahedron are all right angles: the new tetrahedron therefore sits very sweetly in one corner of a cuboid XYSZ'RW'QP (Figure 125).

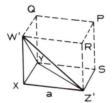

Figure 125

If we denote the length of XY by $b$ and the length of XW' by $c$, then the area A of W'XY is just $\frac{1}{2}b \cdot c$. We have to show that the volume V of the tetrahedron W'XYZ' (and hence the volume of the original tetrahedron WXYZ also) is equal to

$$\tfrac{1}{3}(\tfrac{1}{2}b \cdot c) \cdot a.$$

In other words we have to show that the volume V of the tetrahedron W'XYZ' is precisely one sixth of the volume ($= a \cdot b \cdot c$) of the whole cuboid. This seems simple enough, but it is not quite as obvious as one might think; (you might like to close the book and try it for yourself before reading further).

\*     \*     \*     \*     \*

The tetrahedron W'SYZ' has base SZ'Y congruent to XYZ', and corresponding height W'X precisely equal to that of the tetrahedron W'XYZ'.

The volume of W′SYZ′ is therefore equal to the volume V of W′XYZ′ (Figure 126). The volume of WXYZ′ is therefore *exactly half* the volume of the pyramid W′XYSZ′ with "*a* by *b*" rectangular base XYSZ′.

Figure 126

But the tetrahedron W′XYZ′ has two other right-angled faces, each of which forms precisely half of one of the cuboid's rectangular faces (Figure 127). Thus the tetrahedron W′XYZ′ forms *exactly half* of two other rectangular based pyramids: the pyramid YW′XZ′R with the "*c* by *a*" rectangular base W′XZ′R, and the pyramid Z′W′XYQ with the "*b* by *c*" rectangular base W′XYQ.

 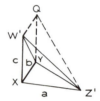

Figure 127

As we have constructed them, the three rectangular-based pyramids W′XYSZ′, YW′XZR, Z′W′XYQ are not separate. But if we take separate copies, then they can be put together to make the complete "*a* by *b* by *c*" cuboid (Exercise 16(ii)). Since the volume of each of these pyramids is precisely twice the volume V of the tetrahedron W′XYZ′, we have our desired conclusion that

$$V = \tfrac{1}{6}(a \cdot b \cdot c).$$

### EXERCISES

1. In our approach to area via the method of inner and outer approximations we measured each shape in terms of *squares*. One reason for using squares is that they make good "tiles": that is, they fit together perfectly, without gaps or overlaps, to cover the whole plane. In 2-dimensions there are many other shapes with this same property (Figure 128). But if a repeating "basic unit" is to be used, then the most valuable alternatives to squares are rectangles and parallelograms.

   In 3-dimensions there are endless possibilities for a repeating "basic unit" beyond the obvious, and most useful, examples of cubes, cuboids and parallelepipeds. But our mathematical experience in 3-dimensions is so weak that we have great difficulty imagining shapes fitting together in space.

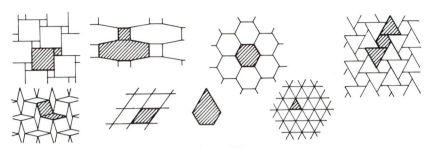

Figure 128

(i) Do regular tetrahedra of a given size make good "bricks"? Could they be fitted together perfectly without gaps and without overlaps to fill space? Would regular octahedra of a given size make good "bricks"? What about regular icosahedra? or regular dodecahedra?

(ii) Suppose we are allowed to combine different kinds of regular polyhedra to make our basic unit. Can one then construct shapes which "fill space" in this way? For example, which of the compound units in Figure 129, each made from one octahedron and two tetrahedra, would fill space? Can you construct other "compound units" like this which fill space?

(iii) Which of the semi-regular polyhedra fill space?

Figure 129

2. Let X be an "$x$ by $y$ by $z$" cuboid. Let $x_i$, $y_i$, $z_i$ be the numbers formed by taking only the first $i$ decimal places of $x$, $y$, $z$ respectively. Let $a_i$ and $A_i$ be the inner and outer approximations to X in terms of "$1/10^i$ by $1/10^i$ by $1/10^i$" cubes. Show that

$$A_i - a_i = \frac{(10^i)^2(x_i\,y_i + y_i\,z_i + z_i\,x_i) + 10^i(x_i + y_i + z_i) + 1}{(10^i)^3}.$$

Hence show that the number $\alpha$ pinpointed by the increasing sequence $a_1, a_2, a_3,$ ..., and the number $A$ pinpointed by the decreasing sequence $A_1, A_2, A_3, \ldots$ are equal.

3. Show that in Exercise 2 we have

$$\alpha = A = x \cdot y \cdot z.$$

4. Reduce an "$x$ by $y$ by $z$" cuboid to a standard "1 by 1 by ?" cuboid. [Hint: One approach is first to reduce the "$x$ by $y$" face to a standard "1 by ?" rectangle as in Chapter III.3, and so to obtain a "1 by ? by $z$" cuboid; one can then reduce the "? by $z$" face of the new cuboid by the same method.]

5. Show that the method you used in Exercise 4 reduces the "$x$ by $y$ by $z$" cuboid to a standard "1 by 1 by $xyz$" cuboid.

6. (i) Find a formula for the area of the regular $n$-gon inscribed in a circle of radius $a$. Find a formula for the difference between the area of the regular $n$-gon circumscribed around, and the regular $n$-gon inscribed in this circle of radius $a$. Hence conclude that the area of the circle is $\pi a^2$.

(ii) Inscribe an $n$-gon $Q_1 Q_2 \ldots Q_n$ in the ellipse

$$\frac{x^2}{b^2} + \frac{y^2}{a^2} = 1$$

as follows: first construct the regular $n$-gon $P_1 P_2 \ldots P_n$ in the circle

$$x^2 + y^2 = a^2$$

with vertices

$$P_k = (a \cos (2k\pi/n),\ a \sin (2k\pi/n));$$

then put

$$Q_k = (b \cos (2k\pi/n),\ a \sin (2k\pi/n))$$

(see Figure 130). Show that

$$\text{area } Q_1 Q_2 \ldots Q_n = \frac{b}{a} \times (\text{area } P_1 P_2 \ldots P_n).$$

Conclude that the area of the ellipse is $\pi ab$.

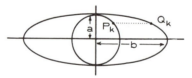

Figure 130

7. Let X be a "finite" solid bounded by six faces which are parallel in pairs. Show that the six faces are all parallelograms and that parallel faces are congruent. What other 3-dimensional figures (bounded by six faces parallel in pairs) are possible if we drop the condition that the figure be "finite?"

8. (i) Show how to reduce any parallelepiped to a cuboid. [Hint: Choose one family of parallel edges and cut perpendicular to these edges.]

(ii) Can every parallelepiped be reduced to a cuboid with one single cut? If so, explain why; if not, give a counterexample.

9. Show that any pyramid with an $(n + 1)$-gon as base can be dissected into $n - 1$ tetrahedra.

10. (i) Just in case you failed to find a proof that, in the plane, every polygon with $n + 1$ vertices can be dissected into an $n$-gon and a triangle (Exercise 13 (ii) of Chapter III.3), here is an outline proof of the stronger fact: that every polygon with $n + 1$ vertices ($n \geq 3$) has *two* disjoint protruding triangles, each of which

can be cut off without interfering with any other edges so as to leave an $n$-gon and a triangle.

*Step 1 :* Check that the assertion is true for all possible $(3 + 1)$-gons.

*Step 2 :* Assume that the assertion is true for all polygons with less than $n + 1$ vertices, and prove that it is then true for any $(n + 1)$-gon as follows:

(a) Given an $(n + 1)$-gon (Figure 131), choose a "protruding vertex" B and call its neighbours A and C.

(b) If the triangle ABC can be safely amputated leaving an $n$-gon, then we must apply our induction assumption to the remaining $n$-gon to show that the original $(n + 1)$-gon also has a second protruding vertex, disjoint from ABC, which can be safely amputated.

(c) If the triangle ABC cannot be safely amputated, other vertices of the $(n + 1)$-gon must intrude into the triangle ABC. We can then dissect the $(n + 1)$-gon into two smaller polygons as follows: Imagine a radial line with centre B, which rotates from BA to BC and stops as soon as it first hits an intruding vertex Z; (if the line hits several intruding vertices all at once, pick the one closest to B). The line segment BZ lies entirely inside the original $(n + 1)$-gon, and dissects it into two smaller polygons. Now apply our induction assumption to these two smaller polygons to show that the original $(n + 1)$-gon necessarily has two disjoint protruding triangles which can be safely amputated.

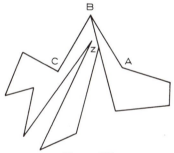

Figure 131

(ii) Let X be a polyhedron with $n + 1$ vertices and P a vertex of X. In order to "cut off" the vertex P and leave a polyhedron with $n$ vertices it is essential that the vertices joined to P *all lie in one plane.* (This is always the case if X is regular, or if each vertex of X has only three neighbours.)

(a) Find a polyhedron X with at least one vertex P such that the vertices joined to P do not all lie in one plane.

(b) Find a polyhedron X, which has $n + 1$ vertices, and which cannot be dissected into a pyramid and a polyhedron with $n$ vertices.

11. (i) Let X be a convex polyhedron whose vertices lie on two parallel faces; that is, X consists of two parallel (convex) polygonal faces A and B with edges running from A to B so as to form a surrounding wall of faces (Figure 132). Show that X can be dissected into a finite number of tetrahedra. Try to get some expression for the number of tetrahedra required in terms of the total number of vertices of the original polyhedron X. [Hint: Try induction on the total number

Figure 132

of vertices. Be careful about starting the induction; one of A and B must have at least three vertices, so you must start by showing how to dissect *any* polyhedron in which A has 1 or 2 vertices and B has 3 vertices (Figure 133) into tetrahedra.]

(ii) Given a convex polyhedron Y with $N$ vertices, fix it in space. Then dissect it by a collection of horizontal planes, one through each vertex, into a collection of polyhedra $X_1, X_2, \ldots, X_M$ of the kind discussed in part (i). Next dissect each $X_i$ into tetrahedra as in part (i). Thus the polyhedron Y has been dissected into a finite number of tetrahedra: *estimate the number of tetrahedral pieces into which Y has been dissected.*

Figure 133

12. (i) You know that the volume of a pyramid with square base of side $b$ and height $h$ is equal to $\frac{1}{3}b^2 \cdot h$. Show that the volume of a truncated pyramid with square base of side $b$, square top of side $a$ and height $h_1$ is $\frac{1}{3}(a^2 + ab + b^2) \cdot h_1$.

(ii) Show that for the truncated pyramid described in (i), the cross-section at height $\frac{1}{2}h_1$ above the base is a square with side $\frac{1}{2}(a + b)$.

13. Let X be a polyhedron whose vertices lie on two parallel faces $A_1$ and $A_2$; such a polyhedron is sometimes called a *prism* and is the 3-dimensional analogue of a trapezium.

(i) The formula $\frac{1}{2}(A_1 + A_2) \cdot h$ for the *area of a trapezium* (Figure 134) can be written in many other ways if one introduces the cross-section $A_3$ midway between the two parallel edges: for example,

$$A_3 \cdot h, \quad \tfrac{1}{2}(A_1 + A_2) \cdot h, \quad \tfrac{1}{3}(A_1 + A_3 + A_2) \cdot h,$$

$$\tfrac{1}{4}(A_1 + 2A_3 + A_2) \cdot h, \quad \tfrac{1}{6}(A_1 + 4A_3 + A_2) \cdot h, \ldots .$$

Show that the *volume of the prism* X is given by

$$\tfrac{1}{6}(A_1 + 4A_3 + A_2) \cdot h$$

where $A_3$ denotes the area of the cross-section midway between the two parallel faces. [Hint: First prove that the formula holds when the total number of vertices of X is at most 5—in which case either $A_1 = 0$ and $A_3 = \frac{1}{4}A_2$, or $A_2 = 0$ and $A_3 = \frac{1}{4}A_1$. Then use induction on the total number of vertices of X.]

(ii) Explain why there can be no general formula for the volume of a prism in terms of $A_1$, $A_2$ and $h$ only.

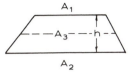

Figure 134

14. Place a circle of radius $r$ on the line $y = -r$ at the point $(-\pi r, -r)$ (Figure 135). Mark the point A on the circle where it touches the line $y = -r$. The circle rolls along the line $y = -r$ with constant speed $r$; as it does so, the point A traces out the curve AA′A″ (Figure 136) which rises from $(-\pi r, -r)$ to $(0, r)$ and then falls to $(\pi r, -r)$. This curve is called the *cycloid*.

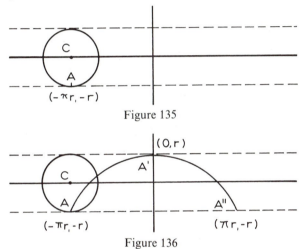

Figure 135

Figure 136

(i) The centre C of the rolling circle starts out, at time $t = 0$, at the point $(-\pi r, 0)$. Find its coordinates at time $t$.

(ii) The point A starts out, at time $t = 0$, at the point $(-\pi r, -r)$. Find its coordinates at time $t$.

(iii) Let $\bar{A}$ be the "projection" of A onto the vertical diameter through C (Figure 137). Observe that the segment A$\bar{A}$ is precisely the cross-section of our basic semi-circle UCVA at the level $y = -\cos t$. Conclude that the coordinates of $\bar{A}$ at time $t$ are

$$(-r(\pi - t), \ -r \cdot \cos t).$$

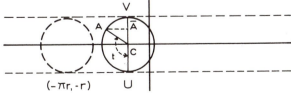

Figure 137

(iv) Sketch the curve $\gamma$ traced out by $\bar{A}$ as $t$ runs from 0 to $\pi$ (Figure 138). Show that $\gamma$ is mapped onto itself by a rotation through $\pi$ about the point $M = (-\pi r/2, 0)$. Conclude that $\gamma$ dissects the rectangle $A_0\, BA_\pi\, D$ into two equal halves, and that the area under the cycloidal arc $A_0\, A_\pi$ is therefore equal to

$\frac{1}{2}$(area of rectangle $A_0\, BA_\pi\, D$) + (area between $\gamma$ and cycloid $A_0\, A_\pi$)

$\qquad = \pi r^2 \qquad\qquad$ + (area between $\gamma$ and cycloid $A_0\, A_\pi$).

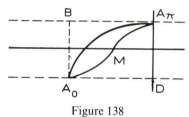

Figure 138

(v) Now use the observation about $A\bar{A}$ in part (iii) to show the following:

at each level $y = -(\cos t)$, the region between $\gamma$ and the cycloidal arc $A_0\, A_\pi$ has exactly the same cross-section $A\bar{A}$ as the semicircle $BC_0\, A_0\, X$ (Figure 139).

(vi) By appealing to the method of slices, conclude that the area under the complete cycloidal arch $A_0\, A_\pi\, A_{2\pi}$ is precisely $3\pi r^2$.

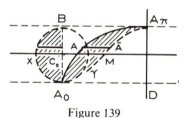

Figure 139

15. (i) Use integration by parts to show that

$$\int_{-\pi/2}^{\pi/2} \cos^{n+1}\theta \; d\theta = \frac{n}{n+1} \int_{-\pi/2}^{\pi/2} \cos^{n-1}\theta \; d\theta.$$

Hence express

$$\int_{-\pi/2}^{\pi/2} \cos^{2n+1}\theta \; d\theta \quad \text{and} \quad \int_{-\pi/2}^{\pi/2} \cos^{2n}\theta \; d\theta$$

in terms of $n$.

(ii) The constants

$$c_2 = \frac{\text{area of circle}}{\text{square of radius}} = \pi,$$

$$c_3 = \frac{\text{volume of sphere}}{\text{cube of radius}} = \frac{4}{3}\pi,$$

form part of an endless sequence whose general term is

$$c_n = \frac{\text{``volume'' of sphere in } n\text{-dimensions}}{n^{\text{th}} \text{ power of radius}}.$$

(a) Let

$$x_1^2 + x_2^2 + x_3^2 + x_4^2 = r^2$$

be a "sphere" of radius $r$ in 4-dimensions. Show that the cross-section perpendicular to the $x_4$-axis at the fixed height $x_4$ is the 3-dimensional sphere

$$x_1^2 + x_2^2 + x_3^2 = (r^2 - x_4^2).$$

Hence calculate the 4-dimensional "volume" of the 4-dimensional sphere, and find $c_4$.

(b) In general show that the $(n + 1)$-dimensional "volume" of a "sphere" in $(n + 1)$-dimensions is given by

$$\int_{-r}^{r} c_n (r^2 - x^2)^{n/2} \, dx.$$

Hence find expressions for

$$\frac{c_{2n+1}}{c_{2n}} \quad \text{and} \quad \frac{c_{2n}}{c_{2n-1}}$$

and calculate

$$c_1, \quad c_2, \quad c_3, \quad c_4, \quad c_5, \quad c_6, \quad c_7, \ldots, \quad c_{2n}, \quad c_{2n+1}, \ldots \, .$$

16. (i) Let WXYZ and WXYZ′ be two tetrahedra on the same base WXY and with equal heights (Figure 140). Show that, for each value of $h$, the two tetrahedra have *congruent* cross-sections at height $h$ above the base.

(ii) Show that an "$a$ by $b$ by $c$" cuboid XYSZ′RW′QR (with XZ′ = $a$, XY = $b$, XW′ = $c$) can be dissected into three rectangular-based pyramids: one on an "$a$ by $b$" base, one on a "$b$ by $c$" base, and one on a "$c$ by $a$" base.

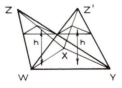

Figure 140

17. Use the method of slices to discover the volume of the solid formed by the intersection of two circular cylinders of radius $a$ whose axes cut at right angles. [Hint: Compare the cross-sections of the solid so defined with those of the sphere inscribed in it—see Problem 25 in Ross Honsberger's lovely book *Mathematical Morsels*, published by the Mathematical Association of America (1978).]

# Curves and Surfaces

Like rulers themselves, length and distance are essentially "*straight*," 1-dimensional concepts; similarly area is an essentially "*flat*," 2-dimensional concept. Yet both concepts are used in other contexts: we do not restrict our attention solely to lengths of straight line segments, but are also interested in "*lengths of curves*," such as the circle, or the cycloidal arch (Exercise 1); and we refer to "*the area*" not just of flat 2-dimensional shapes, but also of curved surfaces such as the cone, cylinder,[1] and sphere. We shall end Part III by inquiring a little more closely how these extended notions of length and area can be justified, though we shall not come to any final conclusion.

If I ask what exactly you mean by "*the length*" of the curve $\gamma$ in Figure 141, you might answer in one of the following ways.

**First Answer:** If the curve $\gamma$ were part of the graph of a function $y = f(x)$ between $x = a$ (at A) and $x = b$ (at B), then the length $l(\gamma)$ of $\gamma$ would be given by the familiar integral formula

$$l(\gamma) = \int_a^b \sqrt{1 + f'(x)^2} \, dx.$$

However, since the curve $\gamma$ wanders about as it does, it could not be the graph of a simple function, but must instead be given in parametrised form $x = x(t)$, $y = y(t)$, where $a \leq t \leq b$ and $A = (x(a), y(a))$, $B = (x(b), y(b))$; the

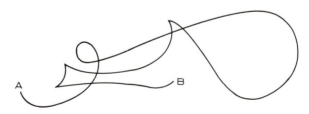

Figure 141

---

[1] Admittedly, the cone and cylinder are "almost flat" in the sense that a single cut allows us to lay each surface flat on the plane. This fact gives us a fairly convincing way of assigning an area to these curved surfaces.

length $l(\gamma)$ of $\gamma$ is therefore given by

$$l(\gamma) = \int_a^b \sqrt{x'(t)^2 + y'(t)^2}\; dt.$$

This answer openly assumes that the curve $\gamma$ is given in terms of nice differentiable functions $x(t)$, $y(t)$, so the same method could not be used to measure the lengths of curves which are given in terms of functions $x(t)$, $y(t)$ which are not differentiable.[2] More remarkable still is the fact that the whole answer seeks to explain something as intuitively elementary as "*the length of a curve*," in terms of differentiation and integration. Do we really need the calculus to explain what we mean by the length of a curve? You might reply that the calculus only provides a convenient way of calculating the length of a curve, but that the *meaning* of the expression "the length of a curve" can be explained without reference to the calculus. Which brings us to your second possible answer to the question—"What do you mean by the length of the curve $\gamma$?"

**Second Answer:** A crude estimate $l_k$ for the length of any curve $\gamma$ is obtained by approximating it by a polygonal arc

$$\mathbf{p}_k = (P_0, P_1, P_2, \ldots, P_k)$$

which has $k$ straight line segments $P_{i-1}P_i$ joining $P_0 = A$ to $P_k = B$ (Figure 142). The length $l_k$ of $\mathbf{p}_k$ is then given by simply adding up the lengths of all the line segments.

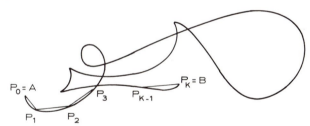

$$P_0 = A \qquad P_3 \qquad P_{K-1} \qquad P_k = B$$
$$P_1 \qquad P_2$$

Figure 142

$$l_k = l(\mathbf{p}_k) = \sum_{i=1}^{k} P_{i-1}P_i.$$

Thus, for example, if $k = 1$, then $\mathbf{p}_1 = (A, B)$ and $l_1 = AB$. As we increase the number of line segments and decrease the length of each line segment, we obtain an endless sequence

$$l_1, \quad l_2, \quad l_3, \quad l_4, \quad l_5, \ldots$$

---

[2] Do such curves exist? If so, can such kinky curves be said to have a well defined length? We can hardly do justice to such questions here, but see Exercise 2 below and Chapter IV.2.

of "approximations" to the length of $\gamma$. *If this endless sequence pinpoints*
*some real number l, then l is called the length of the curve $\gamma$.*

This answer does provide a sort of method for approximating the length
of $\gamma$, which may in fact work quite well in certain particular instances
(Exercise 3). But for us, the real advantage of this approach is that it
apparently makes no appeal to the calculus (though it cannot avoid using
infinite processes). So let us try to reformulate the approach as carefully as
we can.

First of all, one obviously cannot use any old sequence of polygonal
approximations: for example, each polygonal approximation to the circle of
radius $r$ in the endless sequence of Figure 143 has total length $8r$. You

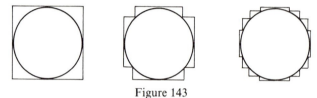

Figure 143

might object that you intended all the vertices $P_i$ of the polygonal approximation to *lie on the curve.* Indeed it is probably wise to make some such
restriction initially—though one gets a hint that it is not quite the right
restriction from the fact that it is not only the endless sequence of inscribed
regular $n$-gons, whose vertices *lie on the curve,* which pinpoint the length of
the circle (Figure 144), but also the endless sequence of circumscribed regular $n$-gons, none of whose vertices lie on the circle (Figure 145). However,
let us insist for the moment that

(i)  the vertices $P_0(=A)$, $P_1$, $P_2$, $\ldots$, $P_k(=B)$ of the polygonal arc $\mathbf{p}_k$ should
     lie in their natural order on the curve segment $\gamma$;

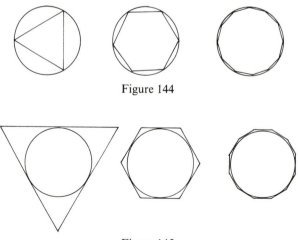

Figure 144

Figure 145

(ii) the total length $l_k$ of $\mathbf{p}_k$ is defined to be

$$l_k = l(\mathbf{p}_k) = \sum_{i=1}^{k} P_{i-1}P_i,$$

(iii) as $k$ increases the lengths of the line segments $P_{i-1}P_i$ which make up $\mathbf{p}_k$ should "tend to zero": that is, for each value of $k$, let $max_k$ be the length of the longest segment $P_{i-1}P_i$ in $\mathbf{p}_k = (P_0, P_1, \ldots, P_k)$

$$max_k = \max_{1 \leqslant i \leqslant k} P_{i-1}P_i,$$

and insist that the endless sequence

$$max_1, \quad max_2, \quad max_3, \quad max_4, \ldots$$

should pinpoint the real number zero.

If after all this the endless sequence

$$l_1, \quad l_2, \quad l_3, \quad l_4, \ldots$$

pinpoints some real number $l$, then $l$ is called *the length of the curve segment* $\gamma$.

But, as it stands, this is still not enough! For if you and I use this definition *independently* to calculate the length of *one and the same curve segment* $\gamma$, it is highly likely that *you* will choose to work with one sequence $\mathbf{p}_1$, $\mathbf{p}_2$, $\mathbf{p}_3$, ... of polygonal arcs satisfying (i)–(iii), while *I* will choose to work with a totally different sequence $\mathbf{p}'_1$, $\mathbf{p}'_2$, $\mathbf{p}'_3$, ... of polygonal arcs satisfying (i)–(iii). Will we always come up with the same answer for the length of the curve segment $\gamma$?

Thus we have a reasonably good way of *calculating* something which we would like to think is the length of the curve segment, but we have not yet proved that the definition is *unambiguous*.

**Third Answer:** If you felt really smart you could turn this dilemma on its head: that is, you could remove all ambiguity from the *definition*, but the uncertainty would simply be shifted to whether our way of *calculating* actually calculates what the definition defines! The details may be found in Exercise 4.


Let us now turn to the second theme of this chapter and ask: *What exactly do we mean by the area of a curved surface* $\Sigma$ *in 3-dimensions?* We defined the length of a curve segment $\gamma$ by considering approximations to $\gamma$ by polygonal arcs $\mathbf{p}_k$—the total length of the polygonal arc $\mathbf{p}_k$ being just the sum of the lengths of its constituent line segments. Since we already know about the areas of plane polygons, it seems reasonable to try to define the area of the curved surface $\Sigma$ by considering approximations to $\Sigma$ by polyhedral surfaces $\mathbf{P}_k$—the total area of the polyhedral surface $\mathbf{P}_k$ being just the sum of the areas of its constituent faces. The conditions which we placed on the sequence of polygonal arcs $\mathbf{p}_k$ approximating the curve segment $\gamma$

suggest the following conditions on the sequence of polyhedral surfaces approximating the curved surface $\Sigma$:

(i) the vertices of each polyhedron $\mathbf{P}_k$ should lie on the surface $\Sigma$;

(i') the faces of each polyhedron $\mathbf{P}_k$ must "cover the whole surface $\Sigma$" in some sense (this is not quite as easy to tie down as in the case of the curve segment $\gamma$ where we simply insisted that $P_0 = A$ and $P_k = B$);

(ii) the total area $A_k$ of $\mathbf{P}_k$ is defined to be the sum of the areas of all the polygonal faces of $\mathbf{P}_k$;

(iii) as $k$ increases, the area of the largest face of $\mathbf{P}_k$ should "tend to zero."

One is then tempted to define *the surface area A of the curved surface $\Sigma$* to be precisely the real number pinpointed by the endless sequence of approximations

$$A_1, \quad A_2, \quad A_3, \quad A_4, \ldots$$

when such a number exists. But in Exercise 6 you will construct a sequence of polyhedral approximations to the cylinder of radius 1 and height 1 which satisfy (i)–(iii), but whose total area increases in an uncontrolled way as $k$ increases! You should try Exercise 6 before reading on.

$$* \qquad * \qquad * \qquad * \qquad *$$

The root of this trouble lies in condition (i). This condition, you will recall, was even called into question in the case of curves. Our insistence that the vertices of an approximating polygonal arc should lie on the curve segment $\gamma$ misses the point, in that it makes no mention of the most important constraint—though *in the case of curves*, this most important constraint happens to be a consequence of our condition (i). Unfortunately we generalised condition (i) rather than the constraint which really matters: namely that the polygonal arc $\mathbf{p}_k$ approximating the curve segment $\gamma$ should approximate the curve $\gamma$ both with respect to *position* (that is, as $k$ increases, $\mathbf{p}_k$ should get closer and closer to $\gamma$) *and with respect to direction* (that is, as $k$ increases, the "slope", or derivative, of each approximating line segment should get closer and closer to the "slope", or derivative, of that piece of the curve it is approximating). It is the fact that this constraint is not satisfied, rather than the fact that the vertices of the polygonal arcs $\mathbf{p}_k$ do not all lie on the curve, which explains why such sequences of polygonal "approximations" as those in Figure 146 and Figure 147 do not pinpoint the length of the corresponding curve segment $\gamma$. And it is the fact that this constraint is not satisfied by the polygonal faces of the approximating polyhedral surfaces in Exercise 6 which explains that paradoxical result also. I shall not

Figure 146

Figure 147

try to prove any of this. But you should perhaps note that our attempt to explain what we meant by "*the length of a curve segment*" and "*the area of a curved surface*" without using the ideas of the calculus appears to have failed!

### EXERCISES

1. In Exercise 14 of Chapter III.4 you found the coordinates at time $t$ of the point A on a rolling circle of radius $r$, which traces out the cycloidal arc from $(-\pi r, -r)$, through $(0, r)$, to $(\pi r, -r)$ (Figure 148): namely

$$A = (-r(\pi - t + \sin t), -r \cos t).$$

Use the familiar integral formula

$$\int \sqrt{x'(t)^2 + y'(t)^2}\, dt$$

to find the length of this cycloidal arc (from $t = 0$ to $t = 2\pi$).

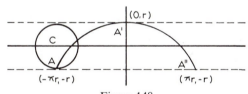

Figure 148

2. (i) Let the "curve" $\gamma_1$ be the square with corners at the points $(\pm\frac{1}{2}, \pm\frac{1}{2})$ (Figure 149). The four line segments making up $\gamma_1$ are all of equal length: divide each into three equal parts. To obtain the curve $\gamma_2$ replace each "1/3 by 1/3" *protruding* corner by the corresponding "1/3 by 1/3" *intruding* corner, and conversely —though at the first stage the curve $\gamma_1$ has no intruding corners (Figure 150). The twelve line segments making up $\gamma_2$ are all of equal length: divide each into three equal parts. To obtain the curve $\gamma_3$ replace each "1/9 by 1/9" protruding corner by the corresponding intruding corner, and conversely.

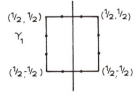

Figure 149

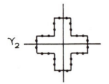

Figure 150

Continuing in this way we obtain an endless sequence of curves

$$\gamma_1, \quad \gamma_2, \quad \gamma_3, \quad \gamma_4, \cdots$$

each of total length 4. Sketch as accurately as you can the curve pinpointed by this endless sequence and find its length.

(ii) Let the curve $\delta_1$ be the square with corners at the points $(\pm\frac{1}{2}, \pm\frac{1}{2})$ (Figure 151). To obtain the curve $\delta_2$ replace each "1/3 by 1/3" protruding corner by the corresponding "1/3 by 1/3" intruding corner (Figure 152). To obtain the curve $\delta_3$ replace each "1/9 by 1/9" protruding corner by the corresponding "1/9 by 1/9" intruding corner—but leave the intruding corners alone (Figure 153).

Continuing in this way we obtain an endless sequence of curves

$$\delta_1, \quad \delta_2, \quad \delta_3, \quad \delta_4, \cdots$$

each of total length 4. Sketch as accurately as you can the curve $\delta$ pinpointed by this endless sequence and find its length.

Figure 151

Figure 152

Figure 153

3. (i) Let $\gamma$ be a circle of radius $r$ and let

$$\mathbf{p}_k = (P_0, P_1, \ldots, P_{k-1}, P_k = P_0)$$

be the inscribed regular $k$-gon. Show that, for $i = 1, 2, \ldots, k$,

$$P_{i-1}P_i = 2r \sin \frac{\pi}{k}$$

and conclude that

$$l_k = l(\mathbf{p}_k) = 2kr \sin \frac{\pi}{k} .$$

Hence show that the endless sequence

$$l_1, \quad l_2, \quad l_3, \quad l_4, \ldots$$

pinpoints the real number $2\pi r$.

(ii) Put $r = 1$ and use a calculator to work out the first twenty terms of the endless sequence

$$\frac{l_1}{2\pi}, \quad \frac{l_2}{2\pi}, \quad \frac{l_3}{2\pi}, \quad \frac{l_4}{2\pi}, \ldots .$$

4. Let $\gamma$ be a curve segment with end points A and B. A sequence $\mathbf{p}_1, \mathbf{p}_2, \mathbf{p}_3, \ldots$ of polygonal arcs

$$\mathbf{p}_k = (P_0 = A, P_1, P_2, \ldots, P_k = B)$$

reaching from $P_0 = A$ to $P_k = B$ is called a sequence of *successive polygonal approximations to* $\gamma$ if

(i) the vertices $P_0, P_1, P_2, \ldots, P_k$ lie on the curve $\gamma$ in natural sequence;
(ii) the vertices of $\mathbf{p}_k$ are always vertices of $\mathbf{p}_{k+1}$ (that is, $\mathbf{p}_{k+1}$ is obtained from $\mathbf{p}_k$ by introducing one extra vertex on the curve $\gamma$ between some pair of vertices of $\mathbf{p}_k$);
(iii) as $k$ increases, the lengths of the line segments $P_{i-1}P_i$ involved in $\mathbf{p}_k$ all "tend to zero".

Let $\mathbf{p}_1, \mathbf{p}_2, \mathbf{p}_3, \ldots$ be such a sequence of successive polygonal approximations to $\gamma$, and let $l_k$ denote the total length of the polygonal approximation $\mathbf{p}_k$. Show that the sequence

$$l_1, \quad l_2, \quad l_3, \quad l_4, \ldots$$

is increasing. Conclude that this sequence either pinpoints a real number $l$, or eventually surpasses each positive whole number $N$; in the latter case we say that the sequence $l_1, l_2, l_3, \ldots$ "pinpoints $+\infty$." Thus each sequence of successive polygonal approximations to $\gamma$ pinpoints either a real number $l$ or $+\infty$.

We can now define unambiguously just what we mean by the *length of the curve segment* $\gamma$. Let $\mathbf{L}$ be the set of all those "numbers" $l$ (where $l$ may be either a real number, or the symbol $+\infty$) which are pinpointed by some sequence $\mathbf{p}_1, \mathbf{p}_2, \mathbf{p}_3, \ldots$ of successive polygonal approximations to $\gamma$. Since $\mathbf{L}$ is not the empty set, it follows from the fundamental property of real numbers (see Exercise 5) that there exists precisely one "number" $\lambda$ (which may be either an ordinary real number, or the symbol $+\infty$) satisfying the two conditions

(1) $\lambda \geq l$ for each "number" $l$ in **L**;
(2) if $x \geq l$ for each "number" $l$ in **L**, then $x \geq \lambda$.

$\lambda$ is called the *supremum* of **L** and is denoted by *sup* **L**. If sup **L** $= +\infty$ we say that the curve segment $\gamma$ *has no length*; if sup **L** is an ordinary real number then we define

$$\text{length of } \gamma = \text{sup } \mathbf{L}.$$

5. Let **L** be any non-empty set of ordinary real numbers (that is, not including $+\infty$).
(i) Show that, if $\lambda_1$ and $\lambda_2$ are "numbers" (either real or $+\infty$) which satisfy conditions (1) and (2) in Exercise 4, then $\lambda_1 = \lambda_2$. Conclude that **L** has *at most one* supremum.

(ii) If we can now show that **L** has *at least one* supremum then it will follow that **L** has precisely one supremum.

*Step 1 :* If there exist no real numbers $x$ satisfying

$$x \geq l \quad \text{for each } l \text{ in } \mathbf{L}$$

show that $\lambda = +\infty$ is a supremum of **L** (check the conditions (1) and (2) in Exercise 4). If there do exist such real numbers $x$, choose one and call it $u_1$; then (since **L** is not empty) choose any number in **L** and call it $l_1$. Check that $l_1 \leq u_1$ (Figure 154).

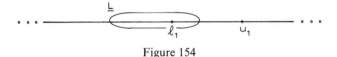

Figure 154

*Step 2 :* Look at the real number $(l_i + u_i)/2$.

(2A): If $(l_i + u_i)/2 \geq l$ for each $l$ in **L** (Figure 155), define new terms $u_{i+1}, l_{i+1}$ as follows:

$$u_{i+1} = \frac{(l_i + u_i)}{2}, \qquad l_{i+1} = l_i .$$

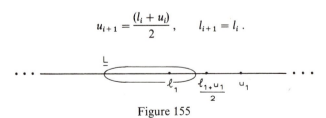

Figure 155

Then repeat Step 2 with $l_{i+1}, u_{i+1}$ in place of $l_i, u_i$.

(2B): If $(l_i + u_i)/2 < l$ for some $l$ in **L** (Figure 156), choose one such number $l$ and call it $l_{i+1}$; let $u_{i+1} = u_i$. Then repeat Step 2 with $l_{i+1}, u_{i+1}$ in place of $l_i, u_i$.

Figure 156

Show that this procedure generates either $\lambda = +\infty$ as a supremum for L as in Step 1, or two endless sequences of real numbers—one increasing

$$l_1, \quad l_2, \quad l_3, \quad l_4, \dots$$

and one decreasing

$$u_1, \quad u_2, \quad u_3, \quad u_4, \dots$$

with

$$l_i \le u_i \quad \text{for each } i,$$

and

$$0 \le u_{i+1} - l_{i+1} \le \tfrac{1}{2}(u_i - l_i).$$

Show that both endless sequences pinpoint one and the same real number; if this real number is called $\lambda$, show that $\lambda$ is a supremum of L.

6. Consider the *lateral* surface of a circular cylinder of radius 1 and height 1 (the top and bottom circular faces have been removed to simplify the problem slightly). Carry out the calculations involved in the following sequence of polyhedral approximations to this surface.

(i) Define the polyhedron $\mathbf{P}_{m,n}$ as follows. First dissect the surface of the cylinder by planes perpendicular to its axis at heights $0/m$, $1/m$, $2/m$, $3/m$, $\dots$, $m/m$ above the base (Figure 157). Then inscribe a regular $n$-gon in each of the resulting $m + 1$ circular sections, in such a way that $n$-gons inscribed in neighbouring circles are rotated through $\pi/n$ relative to one another (Figure 158). Finally construct the faces of the polyhedron $\mathbf{P}_{m,n}$ by completing each isosceles triangle whose base is an edge of some regular $n$-gon and whose apex lies just above or below the base on a neighbouring circle ($\mathbf{P}_{3,5}$ is shown in Figure 159).

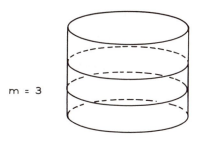

$m = 3$

Figure 157

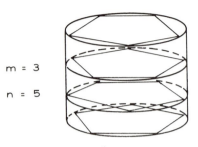

$m = 3$

$n = 5$

Figure 158

Figure 159

(ii) Let $m = m(n)$ be a function of $n$ (say $m = $ constant, or $m = $ nearest integer to $\sqrt{n}$, or $m = $ nearest integer to $\frac{1}{2}n$, or $m = n$, or $m = 2n$, or $m = n^2$, or ...). Check that the endless sequence of polyhedra

$$\mathbf{P}_{m(3),\,3}, \quad \mathbf{P}_{m(4),\,4}, \quad \mathbf{P}_{m(5),\,5}, \quad \mathbf{P}_{m(6),\,6}, \cdots$$

approximates the surface of the cylinder in the sense of conditions (i)–(iii) on page 236.

(iii) Work out a formula for the area $A_n$ of $\mathbf{P}_{m(n),\,n}$, and show that the "number" pinpointed by the endless sequence

$$A_3, \quad A_4, \quad A_5, \quad A_6, \ldots$$

depends on the number pinpointed by the endless sequence

$$\frac{3}{m(3)}, \quad \frac{4}{m(4)}, \quad \frac{5}{m(5)}, \quad \frac{6}{m(6)}, \ldots$$

[Hint: The formula for $A_n$ may involve awkward-looking factors such as $(\sin(\pi/n))/(\pi/n)$, or $(\sin(\pi/2n))/(\pi/2n)$, but these rapidly approach 1 as $n$ increases.]

(iv) Choose the function $m(n)$ so that the endless sequence

$$A_3, \quad A_4, \quad A_5, \quad A_6, \ldots$$

pinpoints the number $2\pi$.

(v) Choose another function $m(n)$ so that the endless sequence

$$A_3, \quad A_4, \quad A_5, \quad A_6, \ldots$$

"pinpoints $+\infty$".

(vi) For which real numbers $\alpha$ can one choose a function $m(n)$ so that the endless sequence

$$A_3, \quad A_4, \quad A_5, \quad A_6, \ldots$$

pinpoints $\alpha$?

PART IV

# FUNCTIONS

In which
- we trace the gradual development of the function-concept from its origins in the purely geometrical and physical problems of the early seventeenth century, up to the eventual emergence of the modern function-concept in the nineteenth century;
- we distinguish between the geometric idea and the algebraic idea of a function, examine their respective inadequacies (and strengths), and see how the fruitful interaction between them eventually made up for their individual weaknesses—giving rise to the modern abstract function-concept;
- and we end with a detailed discussion of one particular class of functions—the exponential functions.

CHAPTER IV.1

# What Is a Number?

In Chapter II.5 we stressed the distinction between

the *intuitive idea* that the counting process is *endless*,

and

the *actual fact* that we have English names for *only finitely many* of these counting numbers.

Fortunately the Hindu-Arabic numeral system allows us to accept

9181716151413121110198765432 1 0

as a number, whether or not we happen to know a colloquial English name for it.[1] Yet we scarcely even notice this curious phenomenon—thanks to the invention and consistent use of a remarkably flexible way of writing numbers in terms of just *ten* basic symbols

1, 2, 3, 4, 5, 6, 7, 8, 9 and—most crucial of all—0.

The notation makes it possible to imagine and to manipulate arbitrary numbers (as in the basic algorithms for carrying out addition, subtraction, multiplication, etc.) no matter how large they may be. Indeed, we become so used to this phenomenon that it comes as something of a shock when we come across a mathematical problem which defies our attempts to find a tidy solution, even though it clearly *ought to have* a nice numerical answer.

We met one such problem in Chapter I.2. It is easy to see that, if $N < 2^n$, then

$$\sum_{r=1}^{N} \frac{1}{r^2} \leq \sum_{r=1}^{2^n-1} \frac{1}{r^2}$$

$$= \frac{1}{1^2} + \left(\frac{1}{2^2} + \frac{1}{3^2}\right) + \left(\frac{1}{4^2} + \frac{1}{5^2} + \frac{1}{6^2} + \frac{1}{7^2}\right) + \dots$$

$$+ \left(\frac{1}{(2^{n-1})^2} + \dots + \frac{1}{(2^n - 1)^2}\right)$$

---

[1] Edward Kasner and James Newman explore similar matters in a very lively section of their classic popularisation: *Mathematics and the Imagination* (Chapter 1: "New names for old", pp. 19–26; and Chapter 2: "Beyond the Googol", pp. 27–35).

$$\leq \frac{1}{1^2} + \left(\frac{1}{2^2} + \frac{1}{2^2}\right) + \left(\frac{1}{4^2} + \frac{1}{4^2} + \frac{1}{4^2} + \frac{1}{4^2}\right) + \ldots + \left(\frac{1}{(2^{n-1})^2} + \ldots + \frac{1}{(2^{n-1})^2}\right)$$

$$= 1 + \quad \frac{2}{2^2} \quad + \quad \frac{4}{4^2} \quad + \ldots + \quad \frac{2^{n-1}}{(2^{n-1})^2}$$

$$= 1 + \quad \frac{1}{2} \quad + \quad \frac{1}{4} \quad + \ldots + \quad \frac{1}{2^{n-1}}$$

$$< 2.$$

Thus it seems obvious that the *endless* sum

$$\sum_{r=1}^{\infty} \frac{1}{r^2}$$

can be assigned some numerical value between 1 and 2—though we can only be sure of this by applying the fundamental property of real numbers to the endless sequence of *finite* sums

$$\frac{1}{1^2}, \quad \frac{1}{1^2} + \frac{1}{2^2}, \quad \frac{1}{1^2} + \frac{1}{2^2} + \frac{1}{3^2}, \quad \frac{1}{1^2} + \frac{1}{2^2} + \frac{1}{3^2} + \frac{1}{4^2}, \ldots$$

(the terms of this sequence are clearly increasing, and each term is less than 2 as we have just proved). Around 1700 Jakob Bernoulli (1654–1705), a Swiss mathematician who had succeeded in assigning numerical values to a number of other endless sums, but who had failed in the case of $\sum(1/r^2)$, wrote

> "If somebody should succeed in finding what till now withstood our efforts and communicate it to us, we shall be much obliged to him."

Now the endless sum

$$\sum_{r=1}^{\infty} \frac{1}{r^2}$$

is in some sense so "natural" that we almost *expect* it to have a tidy, compact value. And in this instance our expectation is justified, though it took the genius of Euler to discover the very satisfying result

$$\sum_{r=1}^{\infty} \frac{1}{r^2} = \frac{\pi^2}{6} .$$

However we should not allow this stroke of luck to deceive us into thinking that every numerical expression which seems to be mathematically "*natural*" can be assigned so satisfying and familar a value as this—as one can see from the following example, which is also due to Euler. It is very easy to show that the numerical value of the ordinary finite sum

$$\sum_{r=1}^{N} \frac{1}{r} = 1 + \frac{1}{2} + \frac{1}{3} + \frac{1}{4} + \ldots + \frac{1}{N}$$

can be made *as large as we please* simply by choosing $N$ sufficiently large (Exercise 1(i)). Thus the endless sequence of finite sums

$$1, \quad 1 + \frac{1}{2}, \quad 1 + \frac{1}{2} + \frac{1}{3}, \quad 1 + \frac{1}{2} + \frac{1}{3} + \frac{1}{4}, \quad 1 + \frac{1}{2} + \frac{1}{3} + \frac{1}{4} + \frac{1}{5}, \ldots$$

increases *without bound*, and so *does not pinpoint any real number at all*. But this does not mean that

$$\sum_{r=1}^{N} \frac{1}{r}$$

is not interesting! For this finite sum is very closely related to the *natural logarithm* of $N + 1$

$$\ln(N + 1) = \int_{1}^{N+1} \frac{1}{r} \, dr.$$

To demonstrate the relationship strictly, we would need a careful definition of the integral—and that would require another book! We simply note that a strict proof of the relationship between

$$\sum_{r=1}^{N} \frac{1}{r} \quad \text{and} \quad \ln(N + 1)$$

would require us to make the following *intuitive* ideas *precise*:

(1) $\displaystyle\int_{1}^{N+1} \frac{1}{r} \, dr$   is equal to the shaded area under the curve

$$y = \frac{1}{r} \quad (Figure \ 160);$$

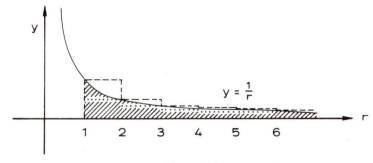

Figure 160

(2) $\sum\limits_{r=1}^{N} \dfrac{1}{r}$ is equal to the area below the dashed upper staircase in *Figure 160;*

(3) $\sum\limits_{r=2}^{N+1} \dfrac{1}{r}$ is equal to the area below the dotted lower staircase in *Figure 160.*

It then follows that

$$\sum_{r=2}^{N+1} \frac{1}{r} < \int_{1}^{N+1} \frac{1}{r}\, dr < \sum_{r=1}^{N} \frac{1}{r}.$$

Hence (Exercise 1(ii))

$$0 < \sum_{r=1}^{N} \frac{1}{r} - \ln(N+1) < 1.$$

Euler proved that, if we define

$$a_{N\cdot} = \sum_{r=1}^{N} \frac{1}{r} - \ln(N+1)$$

(so that each $a_N$ lies between 0 and 1), then the endless sequence

$$a_1, \quad a_2, \quad a_3, \quad a_4, \quad a_5, \ldots$$

does not simply wander around aimlessly between 0 and 1 but *definitely pinpoints some number* $\gamma$. But although the definition of this number is very "natural" in some sense, and although one can calculate the first few decimal places (Exercise 2)

$$\gamma = .5772156649 \ldots,$$

no satisfactorily tidy alternative description of $\gamma$ is known. It is not even known whether $\gamma$ is rational or irrational!

But one does not need to venture forth into higher mathematics for examples of elusive numbers. Consider, for example, the following well-known problem!

*A goat is tethered by a rope[2] from a post on the boundary of a circular field of unit radius. What length r of rope will allow the goat access to precisely half the field?*

At first sight this would seem to be a fairly ordinary geometrical problem, and an initial diagram (Figure 161) shows clearly that $r$ is greater than 1, and considerably less than 2. The obvious approach reduces the problem to the solution of a very simple-looking equation (you should try the problem

---

[2] The goat is a "point-goat," and the rope is "inextensible."

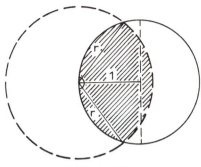

Figure 161

for yourself before looking at Exercise 3). But unfortunately it is not at all clear how to solve it.

The original problem seemed perfectly straightforward; the number $r$ certainly exists and should, one feels, have some "*natural*" interpretation: yet it turns out to be irritatingly elusive.

Now the fundamental objects studied by the calculus, and by the whole of modern mathematics, are *functions* rather than *numbers*. But we shall soon see that functions confront us with difficulties which are very closely analogous to, if more far-reaching and more complex than, those which arise in the case of numbers. Moreover, our notation for functions is nowhere near as flexible as our decimal notation for real numbers. As a result, all the potential difficulties which were raised in Chapter II.5, and which, in the context of numbers, might have appeared over-exaggerated, become clearly visible in the context of our own understanding of functions.

EXERCISES

1. (i) Show that

$$\frac{1}{N+1} + \frac{1}{N+2} + \dots + \frac{1}{2N} \geq \frac{1}{2}.$$

Conclude that

$$\sum_{r=1}^{2N} \frac{1}{r} \geq \sum_{r=1}^{N} \frac{1}{r} + \frac{1}{2}.$$

Hence show that

$$\sum_{r=1}^{2^n} \frac{1}{r} \geq 1 + \frac{n}{2}.$$

(ii) Use the inequalities

$$\sum_{r=2}^{N+1} \frac{1}{r} < \int_1^{N+1} \frac{1}{r}\,dr < \sum_{r=1}^{N} \frac{1}{r}$$

to show that

$$0 < \sum_{r=1}^{N} \frac{1}{r} - \ln(N+1) < 1.$$

2. Let

$$a_n = \sum_{r=1}^{n} \frac{1}{r} - \ln(n).$$

(i) Investigate the behaviour of the endless sequence

$$a_1, \quad a_2, \quad a_3, \quad a_4, \quad a_5, \dots .$$

[For example, procure a programmable calculator, and write a program which does something like the following:

"Label": Increase the contents of 'memory 1' by 1. ('memory 1' contains 'N').

Recall contents of 'memory 1' and 'pause'. ('N' becomes briefly visible in display).

Take the reciprocal (1/x) of this number.

Add result to the contents of 'memory 2'. ('memory 2' contains '$\sum_{r=1}^{N} \frac{1}{r}$').

Recall contents of 'memory 2' and 'pause'. ('$\sum_{r=1}^{N} \frac{1}{r}$' briefly visible in display).

Recall contents of 'memory 1'. Take its natural logarithm (ln x).

Change the sign (+/−).

Add result to (*recalled*) contents of 'memory 2' and 'pause'. ('$a_N$' briefly visible in display).

GOTO "Label"

It is instructive to watch the display as $N$ increases, looking for apparently uniform behaviour: for example, the terms decrease—surprisingly quickly at first, ($a_1 = 1$, $a_2 = .806$ ..., $a_3 = .734$ ..., $a_4 = .697$ ..., ...), but much more slowly later, ($a_{10} = .626$ ..., $a_{20} = .602$ ..., $a_{40} = .589$ ..., $a_{100} = .582$ ..., $a_{638} = .577$ ..., ...).]

(ii) Prove that, for each $n$, $a_n > a_{n+1}$ provided that $1 < \ln[(n+1)/n]^{n+1}$.

(iii) Use the definition of $e$ given in Chapter II.10, Exercise 1(ii), to prove that, for each $n$, $e < [(n+1)/n]^{n+1}$.

(iv) Now use the fundamental property of real numbers to conclude that the endless sequence $a_1, a_2, a_3, \dots$ pinpoints some number $\gamma \geq 0$. Can you improve on the inequality $\gamma \geq 0$?

3. A goat is tethered by a rope of length $r$ from a post on the boundary of a circular field of unit radius; the length $r$ is such that the goat has access to *exactly half* the field.

(i) Show that $r = 2 \cos(x/2)$ (where $x$ is the angle shown in Figure 161).

(ii) Show that $x$, and hence $r$, is given by the equation

$$\sin x - x \cos x = \frac{\pi}{2}.$$

# What Is a Function?

> Nobody can explain what a *function* is, but this is what really
> matters in mathematics: "A function *f* is given whenever with
> every real number *a* there is associated a number *b* (as for
> example, by the formula $b = 2a + 1$). *b* is then said to be the
> value of the function *f* for the argument value *a*."
> Hermann Weyl, *Philosophy of Mathematics*
> *and Natural Science*

When we asked the question

*What is a number?*

we certainly did not expect an answer such as

*1 and π are numbers.*

Nevertheless, the way in which we learn as children what is meant by "*a
number*" depends on our experiencing and exploring precisely such *specific
examples* as these. The jump from such *local* experiences of specific numbers, to a suitably *global* number-concept is greatly assisted by two remarkably flexible and fruitful mental images: first

(i) *the decimal notation* for whole numbers, which extends first to
decimal fractions, and later to endless decimals;

and second

(ii) the idea of *the number line.*

Now the remarkable thing about these mental images is that they give an
essentially accurate picture of the structure and the behaviour of abstract
real numbers, and so allow us to come to grips with the general concept of
"*a real number*" in a more or less explicit way.

Our early experience of *functions* is in many ways similar to that of (real)
numbers: that is, we first get to know and to represent particular examples
such as $n^2$ and $2^n$. But the transition from such particular examples to a
suitably general function-concept is far from automatic.[1] Indeed, to provide
a rough and ready framework for the ensuing discussion it does not seem to
me unreasonable to suggest that *the difficulty inherent in making this transition* from specific examples of what we would now call "functions" to an

adequately general function-concept *was one of the main obstacles in the way of explaining precisely why, when and how the methods of the calculus could be trusted.* Though this is a very crude over-simplification, I shall try to explain why it seems to me a potentially fruitful one—at least for the student who has learned to *use* "naive calculus," and who is looking for an initial appreciation of its limitations.

The problems of finding *tangents* and *areas*, which gave birth to the calculus, concerned *curves* rather than functions. But with the introduction (around 1640) of *the method of coordinates* it became possible to reinterpret these

> *geometric problems about curves*

into the language of

> *algebraic relations between "variable coordinates x and y."*

The natural geometric interpretation of the original problems concerning tangents and areas had generated some brilliant *ad hoc* solutions—from Archimedes in ancient Greece, and from a number of mathematicians in the late sixteenth and early seventeenth centuries. But the geometric approach failed to reveal the extent of the *formal similarities* between apparently unrelated problems. Admittedly there were strong hints that underlying similarities were waiting to be uncovered—at least in the case of area problems, with Cavalieri (1598–1647) publishing his famous "Principle"[2] as Theorem 1, Proposition 1 in Book VII of *Geometria indivisibilibus continuorum nova quadam ratione promota* (1635). But these hints did not at that time give rise to any reliable or systematic general methods. However once the underlying *geometric* problems had been translated into the language of *algebraic* symbols there was at least a chance that systematic methods would be developed which might reveal the *formal similarity* between what were superficially distinct problems. Indeed, Descartes showed in his book *La Géometrie* (1637) that, for a curve whose equation is given in the form $y = f(x)$, it is possible (at least when $f(x)$ is a polynomial of small degree) to find the normal, and hence the tangent, at any specified point of the curve.[3] Descartes' method was rather cumbersome, but a more flexible method (involving "small increments in $x$") had been introduced some years earlier by Fermat (1601–1665).

---

[1] In real life, the formation of our ideas about *functions* may be complicated rather than helped by the simultaneous development of related ideas such as *equations* ($3x^2 + 2x = 5$, $3x^2 + 2xy = 5$), *algebraic expressions* $((5 - 3x^2)/2xy)$, *algebraic identities* ($3x^2 + 2xy = 3x(x + \frac{2}{3}y)$), *formulae* and *graphs* (of $3x^2 + 2xy = 5$, or of $y = (5 - 3x^2)/2x$). However these possible complications will not be discussed here.

[2] Cavalieri's Principle is essentially what we called "the method of slices" in Chapter III.4.

[3] Pages 95–112 in the English translation: *The Geometry of René Descartes*, Dover (1954).

It was from such work as this that Newton and Leibniz later developed their respective *calculi*,[4] which in a sense "explained" why all tangent problems (and all area problems) exhibited a formal similarity, and which demonstrated the inverse relationship between tangent and area problems. At first these *calculi* applied not to functions but to "*variables*" $x$, $y$, via their "*infinitesimal increments*" $dx$, $dy$. Moreover, the rules for manipulating these "variables" and their "infinitesimal increments" were, strictly speaking, self contradictory. Infinitesimal quantities had been used and mistrusted since the time of the ancient Greeks, and nothing had been done to resolve the logical inconsistencies in their use. But as long as the calculi of Newton and Leibniz continued to work so well, there was little incentive to take this fundamental weakness of "infinitesimals" too seriously; neither did there appear to be any urgent need to reconsider the elusive notion of a "variable.

For much of the eighteenth century the fundamental objects studied by the calculus were still thought of as *variables* rather than as functions. And though, little by little, mathematicians came to discuss the calculus in terms of *functions*, their ideas of what constituted a respectable function had to undergo a number of subtle transformations before they eventually arrived at an adequately general function-concept.

The intervening period can be crudely described as "a creative tug-of-war" between two powerful, but contrasting, mental images: first

(i) the idea of a *geometric curve*, or *graph*

and second

(ii) the idea of an *algebraic relation, formula* or *expression* linking two (or more) "variables".[5]

Both images have their roots in the period before the invention of the calculus, and they still have a role to play in helping us get to grips with the

---

[4] The meaning of the word *calculus* was discussed in Chapter I.1.

[5] Just how crude a characterisation this is can be seen from two examples. Firstly, it fails to distinguish between purely algebraic and more general *analytic* conceptions of a function—lumping them together on the grounds that they both consider functions to be given by an expression involving variables and operators. Secondly it makes no explicit reference to the influence of the sciences—ignoring the fact that the function-concept stems in part from ideas about, and experiments involving, measurable causes and effects, and from the study of time dependent phenomena (such as Galileo's study of the distance travelled by a freely falling body).

The pedagogical significance of these mental images is touched on only incidentally in Part IV: students' private perceptions of mathematics seem to be even more difficult to analyse than mathematics itself. For example: the picture associated with a simple quadratic, such as $y = x^2$, is generally interpreted both as the *geometric curve with equation* $y = x^2$ amd as the *graph of the function* $x \mapsto x^2$, whereas in the case of straight lines and linear equations $y = mx + c$ students are often less flexible in that they perceive *the straight line with equation* $y = mx + c$, but not *the graph of the linear function* $x \mapsto mx + c$.

general idea of a function. But these mental pictures are inadequate for our purposes in that they conceal certain crucial features of the currently accepted function–concept.[6] Moreover, images as powerful as these can easily reinforce "*intuitions*" about functions which turn out to be irrelevant—even to the extent where further developments are seriously impeded. We should not therefore be surprised to discover a protracted struggle, during which mathematicians sought to inch their way to a more general function–concept, by stretching first one image, then the other, then the first again, and so on.

The idea of a *geometric curve* strongly suggests something that can be drawn by hand (Figure 162). Implicit in this is the idea that, in any given piece of curve, the *corners* have to be "*reasonably well spaced out.*" The

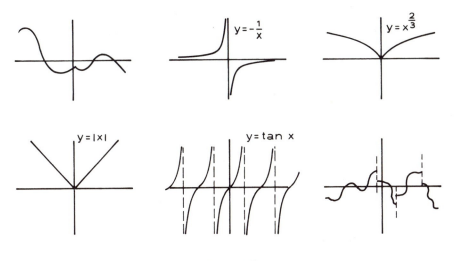

Figure 162

graph of $y = |x|$ in Figure 162 has just one corner (at $x = 0$), but we can certainly construct graphs with infinitely many corners by simply repeating this corner periodically (Figure 163). However, the corners of the resulting graph *remain* reasonably well spaced out. With a little more thought we may even imagine pictures having infinitely many corners between $x = 0$ and $x = 1$ (Figure 164) (though we can never quite draw them!) A second, third, fourth, etc. application of the same idea produces pictures with cor-

[6] One way of counteracting this is to supplement the geometric and algebraic images by a third: namely *mapping diagrams* and *function-boxes* (see, for example *Prof. E. McSquared's Original, Fantastic and Highly Edifying Calculus Primer* by H. Swann and J. Johnson, published by William Kaufmann Inc. (1975)).

Figure 163

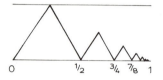          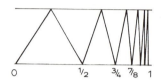

Figure 164

ners galore between $x = 0$ and $x = 1$ (Figure 165) but there is a very real
sense in which even these corners are *reasonably well spaced out*—for if they
were not, would not the picture become almost impossible to imagine as
well as impossible to draw? One *may* nevertheless begin to feel that it is at

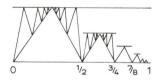

Figure 165

least conceivable that there exist really weird examples in which corners
crop up almost everywhere, but it is not clear how one could construct such
examples on the basis of the geometric image alone. Yet, once one suspects
on *geometric* grounds that such functions exist, it is not all that hard to
construct explicit examples—provided only that one is willing to go back to
the *algebraic* image and look for a suitable expression of the form $y = f(x)$
(Exercise 2).

Now just as we are used to graphs with corners, so we are also familiar
with geometric curves, or graphs (such as $y = -1/x$ and $y = \tan x$) which
have *jumps*, or *discontinuities*, at isolated points.[7] It is then fairly easy to
imagine graphs with infinitely many *jump-points* between $x = 0$ and $x = 1$:
such graphs arise, for example, when one tries to "graph the slope" of the
graphs in Figure 164 (Figure 166). It is not hard to dream up more complicated examples, but the idea of a "*geometric curve*" seems at first to
involve implicitly the assumption that the jump-points too must be

---

[7] For elementary functions the most familiar jumps occur where the graph goes off to $\pm \infty$ (as
in $1/x$ at $x = 0$). But there are other kinds of jumps (as in $e^{1/x}/(1 + e^{1/x})$ at $x = 0$).

"reasonably well spaced out." However, these geometric examples and their obvious extensions may lead one to suspect that perhaps there do exist graphs in which the jump-points crowd more closely together in some sense. But to construct such examples one may have to turn back to the *algebraic* image and look for a suitable expression of the form $y = f(x)$ (Exercise 3).

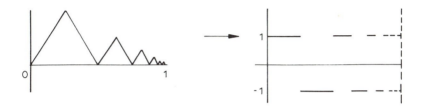

Figure 166

Thus, at a push, the mental image of a *geometric curve*, or *graph*, allows us to imagine *pictures* which are considerably more general than the usual examples of polynomials, rational functions, trigonometric functions, exponential functions (and their inverses). But there are two rather obvious constraints: first of all, one is unlikely to indulge in such gratuitous attempts to extend one's "naive" function-concept unless the naive view has proved itself inadequate in some way; and secondly, even when one has begun to suspect the existence of inconceivably wild *pictures*, the only obvious way of constructing explicit examples *in a form which will allow one to test precisely how wild they are*, is to go back to the *algebraic* image and to exhibit not the picture itself, but rather *an equation* of the form $y = f(x)$, whose graph is the required picture (Exercises 2, 3).

*       *       *       *       *

In the rest of this chapter we shall consider certain important features of the way the function-concept has evolved historically. Our aim is to justify the claim that its evolution can be profitably interpreted as a creative tug-of-war between the geometric and the algebraic images of a function. The actual historical process is too complicated to analyse in detail here, but I have tried to give the flavour of some of the difficulties which had to be overcome. Before embarking on this simplified, but hopefully instructive, historical survey, it may be advisable to reflect consciously on one's own experience of an expanding function-concept by working through Exercise 7 (and 8).

*       *       *       *       *

If the calculus was developed in the *seventeenth* century as a way of solving essentially geometrical problems, and the "variables" involved in calculations usually had obvious geometrical interpretations, then the history of the calculus in the *eighteenth* century may be characterised as a (temporary) triumph of algebra over geometry.[8] As more complicated problems were tackled, the *geometrical meaning* of the variables no longer influenced the methods of solution in the same direct way. The calculus thus became chiefly a way of manipulating algebraic formulae. Euler, for example, described *analysis* as the science of *analytic* (or what we have called *algebraic*) expressions. In his book *Introduction to the Analysis of Infinities* (1748) he stated that

> "a *function* of a variable quantity is an analytical expression composed in whatever way of that variable and of numbers and constant quantities."

Now, as we have suggested above, the *algebraic* image of a function as "an algebraic formula, or expression" has certain advantages over the *geometric* image of a function as "a geometric curve, or graph." For example, however complicated an algebraic expression may be, one can at least begin to *calculate* with it (see Exercises 1, 2, 3). It is in fact generally acknowledged that the transformation of

conceptually hard mathematics

into

routine algebraic manipulation

has a number of advantages *provided one retains some mathematical "feeling" for what the algebraic expression represents and for what is being calculated.*[9]

The most obvious advantage is that the resulting methods become accessible to non-specialists. But there is another advantage which is more immediately relevant to the evolution of the function-concept; that is, the familiar fact that *algebraic symbols have a curious way of insinuating new meanings into old ideas.* When a piece of mathematics is first expressed symbolically, the symbols retain their original meaning; but given time they

---

[8] The same period witnessed the triumph of Leibniz's calculus, with its strongly algebraic flavour, over Newton's, with its geometric and dynamic imagery.

[9] In our haste to cash in on the advantages of algebra—whether in the solution of equations, coordinate geometry, linear algebra or whatever—it is all too easy to forget the precondition in italics. Our "feeling for mathematical meaning" is not part of the *logical* structure of mathematics, but it still needs educating: an algebraic system may be *logically* sufficient, but it all too often remains *psychologically* deficient in meaning. The premature adoption of a formal, algebraic approach simply sorts students into those who manage to discover the underlying meaning for themselves, and those who are reduced to mindless manipulations; some in this latter category are sufficiently clever to master even mindless manipulations, while others remain permanently at sea.

often take on a life of their own. Consider the following fairly typical examples.

(1) The notation $2^2$, $2^3$, etc., for powers gives rise to the "index law"

$$2^m \cdot 2^n = 2^{m+n}$$

for *positive whole numbers m, n*. But once the law has been formulated symbolically like this it is only a matter of time before some bright spark[10] infers a meaning for $2^0$ (since $2^m \cdot 2^0 = 2^m$ suggests that $2^0 = 1$), for $2^{-1}$ (since $2^1 \cdot 2^{-1} = 2^0 = 1$ suggests that $2^{-1} = \frac{1}{2}$), and even for $2^{1/2}$ (since $2^{1/2} \cdot 2^{1/2} = 2^1$ suggests that $2^{1/2} = \sqrt{2}$).

(2) The coefficients in

$$(x + a)^2 = x^2 + 2xa + a^2$$
$$(x + a)^3 = x^3 + 3x^2a + 3xa^2 + a^3$$
$$(x + a)^4 = x^4 + 4x^3a + 6x^2a^2 + 4xa^3 + a^4$$
$$\vdots$$

are the entries in Pascal's triangle. If one expresses these algebraically one obtains the binomial theorem for *positive whole number exponents n*:

$$(x + a)^n = x^n + nx^{n-1}a + \frac{n(n-1)}{2!} x^{n-2}a^2$$
$$+ \frac{n(n-1)(n-2)}{3!} x^{n-3}a^3 + \ldots + a^n.$$

If one then discovers (by long division or some other method) that

$$(1 + a)^{-1} = \frac{1}{1+a} = 1 - a + a^2 - a^3 + \ldots$$
$$= 1 + (-1) \cdot a + \frac{-1 \cdot (-1-1)}{2!} a^2$$
$$+ \frac{-1 \cdot (-1-1) \cdot (-1-2)}{3!} a^3 + \ldots,$$

---

[10] These do not pretend to be strictly historical examples. For example, the compact algebraic notation, which makes it so easy for us to manipulate powers, evolved very slowly and finally emerged long after the ideas of $0^{th}$, negative, fractional, and even irrational exponents had first been suggested. Positive integer exponents and the corresponding index laws occur incidentally in the work of Archimedes, Appolonius and Diophantos (who also considered their reciprocals). $0^{th}$ and negative exponents were used by Nicholas Chuquet in a manuscript entitled *Triparty en la science des nombres* (1484), while positive fractional exponents were used even earlier by Nicole Oresme in *De proportionibus proportionum* (ca. 1360).

then this may encourage the suspicion that the binomial expansion of $(x + a)^n$ is meaningful even when $n$ is *not a positive whole number*.[11]

(3) When Napier (1550–1617) first published details of his "logarithms," he did not strictly use exponents. But very shortly thereafter Henry Briggs (1561–1631) suggested that Napier's invention would be easier to understand and use if 10 was taken as "base," and the logarithm of a number $x$ was defined to be "the exponent of 10 which was equal to $x$": that is

$$x = 10^{\log x}.$$

From this it follows that log $x$ *is only meaningful when $x$ is positive*. Thus, the graph of $y = \log x$ (Figure 167) has a gaping hole to the left of the $y$-axis, which positively invites speculation. Someone who has learned to exploit, *and to trust*, the often unexpected flexibility of the differential calculus, may observe that strict application of the rules of differentiation *regardless of meaning* yields

$$\frac{d}{dx} \log(-x) = \frac{d}{dx} \log(x).$$

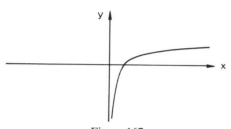

Figure 167

In other words: if it is possible to extend the graph of $y = \log x$ to cover negative values of $x$, then its slope at the point $-x$ is exactly the same as the slope at the point $x$. It is not too hard to see what such a curve would be like (try it for yourself before turning to Exercise 4); unfortunately the curve one draws contradicts one's expectation that

$$\log(-x) = \log(x \cdot -1) = \log x + \log(-1).^{12}$$

Precisely how one gets from the confusion of this last example to complex logarithms remains unclear; but the examples do at least suggest that the speculative habit of manipulating symbols (like $\log(-x)$) in a purely formal

---

[11] Newton's original discovery of the binomial theorem arose in this way. However, his was not idle speculation. He already knew how to "integrate" $x^n$ ($n \neq 1$) and was trying to "integrate" $(1 - x^2)^{1/2}$: that is, he was trying to use his embryonic integral calculus to calculate the area of a circle.

[12] The matter had been considered by both Leibniz and Johann Bernoulli (1667–1748). At different stages of his career Bernoulli supported each of two contradictory positions: firstly $\log(-1) = \pi \cdot \sqrt{-1}$, and secondly $\log(-1) = 0$. Leibniz guessed that $\log(-1)$ is imaginary.

way, regardless of their original meaning, can lead to serious and far-reaching developments in mathematics. However, to be fair, we should perhaps admit that by no means all speculative manipulations of this kind have such memorably happy endings.

We shall end this chapter by examining various aspects of the *algebraic idea of a function* in the period 1772–1837: that is, from the publication in 1772 by Lagrange (1736–1813) of his view that one should analyse functions in terms of their Taylor series expansions, to the work of Dirichlet (1805–1859) on Fourier series, in which the modern idea of a function—as "any well-defined correspondence" *whether or not it can be given by algebraic formulae*—began to be taken seriously.

The algebraic idea of a function had originally been very closely tied to the most familiar examples—the so-called *elementary functions*: that is, polynomials, rational functions, trigonometric functions, exponential functions, and their inverses (roots, inverse trigonometric functions and logarithms). But the accepted meaning of the word *function* was gradually extended, and by the early 1800's one finds what appears to be a measure of agreement as to the correct generalisation of this naive algebraic idea. Thus in the 2nd edition of *Théorie des fonctions analytique* (1813) by Lagrange we read

> "We call *function* of one or several variables any expression of calculation in which these quantities occur in any manner, linked or not with some other quantities that are regarded as having given or constant values, whereas the variables of the function may take all possible values. [ ... ]
> The word *function* was used by the first analysts[13] to denote the powers of one variable. Since then, the meaning of this term has been extended to any quantity formed in any manner from any other quantity."

From the context it is clear that Lagrange understood his words "quantity formed in any manner from any other quantity" to mean that the function be formed *as an algebraical expression* in any manner from any other quantity.

Around the same time, in Chapter 1 of Cauchy's *Cours d'Analyse* (1821), we find

> "When variable quantities are related in such a way that, once one of them is given one can conclude the values of all the others, one usually thinks of these variables as all being expressed in terms of the one, which is called the *independent variable*, and the other quantities expressed in terms of the independent variable are what one calls *functions* of this variable."

Again, though at first sight it is conceivable that the writer might almost be thinking of entirely arbitrary functions, it soon becomes clear that the

---

[13] This seems to refer to Fermat and his contemporaries in the seventeenth century.

quantities are presumed to be related by some collection of *algebraic* (or analytic) expressions.

> "When functions of one or more variables can be [ ... ] immediately expressed in terms of these variables, they are called *explicit functions*. But when one only has the relations between the functions and the variables, that is to say, the equations which these quantities must satisfy, but such that these equations cannot be solved algebraically, then the functions not being immediately expressible in terms of the variables, are called *implicit functions*. In order to make [such functions] explicit it is enough, where possible, to solve the equations which define them."

As these extracts indicate, the algebraic idea of a function as being given by any kind of algebraic equation of formula, had by this time been pushed more or less to its logical conclusion. As far as one can see, no constraints were imposed on the way a function was represented, as long as it was recognizable as an algebraic (or analytical) expression relating the variables in the required deterministic way. In particular endless sums were certainly accepted as particular examples of such algebraic (or analytical) expressions. For example, the representation of functions as *power series* (that is, as *endless* polynomials) had been important from the very first days of Newton's own researches. The importance of the binomial expansion of $(x + a)^n$ when $n$ is no longer a positive whole number, of the power series for $\sin x$, $\cos x$, $e^x$, and of Taylor series in general, had made power series into a basic tool of analysis.

But in spite of this generalisation of the function-concept from elementary functions to arbitrary algebraic expressions, including endless sums, it had been more or less taken for granted that such algebraic (or analytical) expressions behaved as decently as the elementary functions themselves. Now elementary functions (such as $x^2 + 3$, $e^{-x^2}$, $\tan x$, $(x^2 + 1)/(x - 1)$, $\log(1 + x)$, $(1 - x)^{-1/2}$, $\sin^{-1} x$) are particularly easy to work with in that the value of the function is *given by a single analytical expression*, and the function is *defined for precisely those values of x for which the expression makes sense*. Elementary functions can also be differentiated as often as one wants, provided the function is defined, by simply applying the basic rules of the differential calculus.

The idea that this happy state of affairs remained true for functions in general was given considerable credence by

the belief that *every* function could be expanded in a Taylor series,

and

the presumption that *endless* power series behave like polynomials of very large degree.

The view that *power series* are not just a useful tool in analysis, but that they represent the correct generalisation of the naive function-concept is a very attractive one. Power series provide both a natural generalisation of ordinary polynomials and a uniform way of representing elementary functions:

$$x^2 + 3 = 3 + x^2$$

$$e^{-x^2} = 1 + x^2 + \frac{x^4}{2!} + \frac{x^6}{3!} + \frac{x^8}{4!} + \cdots$$

$$\tan x = x + \frac{x^3}{3} + \frac{2x^5}{15} + \frac{17x^7}{315} + \cdots \ ^{14}$$

$$\frac{x^2 + 1}{x - 1} = -(1 + x^2)(1 + x + x^2 + x^3 + \cdots)$$

$$= -(1 + x + 2x^2 + 2x^3 + 2x^4 + \cdots)$$

$$\log(1 + x) = x - \frac{x^2}{2} + \frac{x^3}{3} - \frac{x^4}{4} + \frac{x^5}{5} - \cdots$$

$$(1 - x^2)^{-1/2} = 1 + \frac{1}{2} \cdot x^2 + \frac{1 \cdot 3}{2^2} \cdot \frac{x^4}{2!} + \frac{1 \cdot 3 \cdot 5}{2^3} \cdot \frac{x^6}{3!} + \cdots$$

$$\sin^{-1} x = x + \frac{1}{2} \cdot \frac{x^3}{3 \cdot 1!} + \frac{1 \cdot 3}{2^2} \cdot \frac{x^5}{5 \cdot 2!} + \frac{1 \cdot 3 \cdot 5}{2^3} \cdot \frac{x^7}{7 \cdot 3!} + \cdots .$$

Moreover, if we assume that it is permissible to integrate and differentiate such expressions as though they were polynomials, we clearly obtain other expressions of exactly the same kind. Thus, for example, when faced with the unfamiliar integral[15]

$$\int \frac{e^x}{x} \, dx$$

we can at least express $e^x/x$ as a power series and integrate term by term

$$\int \frac{e^x}{x} \, dx = \int \left( \frac{1}{x} + 1 + \frac{x}{2!} + \frac{x^2}{3!} + \cdots \right) dx$$

$$= \ln(x) + x + \frac{x^2}{2 \cdot 2!} + \frac{x^3}{3 \cdot 3!} + \cdots .$$

And even though there is no obvious way of recognising the function which corresponds to the resulting endless sum, the idea that every function can be represented in this way offers at least the prospect of developing *universally applicable methods for calculating with functions through their power series representations*. More important still from our point of view was the fact that Lagrange sought to exploit his belief that every function could be

---

[14] See Exercise 9.

[15] This example comes from the 2nd edition of Lacroix's *Traité du calcul differentiel et integral* (1810), Vol. III, p. 512.

expanded as Taylor series about each point $x = a$ in order to get round the logical inconsistencies inherent in what was then the accepted definition of the differential or derivative of a function.

Lagrange's dream was hopelessly optimistic. To expand a function $f(x)$ as a Taylor series about the point $x = a$, *we must be able to differentiate the function $f(x)$ as often as we please at the point $x = a$.* But we have already seen that, once one accepts functions defined by endless sums, one can produce bona fide functions (Exercise 3) which are *continuous but not differentiable even once at any point.* In practice Lagrange's vision failed not so much because of such counterexamples as this,[16] but because his approach showed no signs of actually working as he himself had intended: in order to avoid working with infinitesimals and crude notions of "limits" involving 0/0 Lagrange had defined the *derived function* of $f(x)$ to be the coefficient of $h$ in the expansion

$$f(x + h) = f(x) + Ah + Bh^2 + Ch^3 + \ldots;$$

yet he never discovered any satisfactory way of using this ingenious definition to actually calculate derivatives of even moderately complicated functions.

In the end the algebraic idea of a function was replaced, but it was by no means rejected: it ceased to provide the definition of what was meant by the word function, and instead became simply the most convenient means of expressing particular examples. The pressures which brought about this change were numerous, subtle, and cumulative. One common feature of a number of these pressures was that their resolution required *a return*, albeit brief, *to the geometric image* of a function.

For example, one of the ingredients in the dispute about the solution $y = y(x, t)$ of the 1-dimensional wave equation

$$\frac{\partial^2 y}{\partial x^2} = \frac{1}{c^2} \frac{\partial^2 y}{\partial t^2},$$

which was discussed briefly in Chapter I.1, was the question as to which functions $y_0(x) = y(x, 0)$ were admissible as the initial configuration of the string just before being released at time $t = 0$. Now there does not seem to be any physical reason why $y_0(x)$ should be given by a nice algebraic expression: indeed, one feels that there is nothing to stop one starting with the string in any position one likes (Figure 168)—provided only that it does not transgress the assumptions implicit in the derivation of the wave equation itself. And there's the rub, for the fact that $y$ has to satisfy the given differential equation for each value of $x$ and each value of $t$ would seem, in particular, *to forbid functions with corners.* This aspect of the dispute was

---

[16] The first example of a continuous nowhere differentiable function seems to be due to Weierstrass in the 1860's.

Figure 168

never resolved, but it sounded a timely warning of the latent tension between the geometric image of a function, which allows one to imagine
almost any initial configuration for the string, and the algebraic image,
which requires that the initial position be given by a nice algebraic expression.

Perhaps the simplest way to prescribe an initial configuration $y_0(x)$
which is not a simple algebraic function is to splice two or more different
algebraic (or analytical) pieces together. For example, define $y_0(x)$ as in
Figure 169. Such "compound" functions are also relevant to other physical
problems. Fourier's analysis of the diffusion of heat along a 1-dimensional
bar shows how the temperature $v = v(x, t)$ of the point with coordinate $x$ at
time $t$ is determined by the temperature distribution $v_0(x) = v(x, 0)$ at time
$t = 0$. If one simply leaves the bar in a uniformly heated environment, one
expects $v_0(x)$ to be a constant function. If one immerses the end $x = 0$ in the
fire, or in a bucket of ice, for a sufficient time before starting the clock (at
$t = 0$), then one might naively expect $v_0(x)$ to be a linear function—
decreasing if $x = 0$ were in the fire, and increasing if $x = 0$ were in ice. By
applying both heating and cooling devices simultaneously at different
points one can imagine that it must be possible to build more complicated
initial temperature distributions $v_0(x)$ by sticking, or splicing, different linear
functions together; and by using insulation, one feels that it should be
possible to get around the restriction to linear functions. Now the algebraic
idea of a function assumed that the function was given uniformly by a single
formula or equation: its derivative could therefore be calculated by simply
differentiating the whole formula or equation. When confronted by a compound function one could perhaps differentiate each separate algebraic (or
analytical) expression; but this approach is unworkable at the junctions
where the separate pieces, or expressions, are spliced together.

$$y_0(x) = \begin{cases} 0 & \text{if } -\pi \leq x \leq 0 \\ \cos(2x) - 1 & \text{if } 0 \leq x \leq \pi \end{cases}$$

Figure 169

Perhaps the most telling challenge to the contemporary algebraic image
of a function came from Fourier's use of trigonometric series to represent
functions in the form

$$f(x) = \tfrac{1}{2}a_0 + \sum_{n=1}^{\infty} (a_n \cos nx + b_n \sin nx).$$

These appear at first sight to be perfectly reasonable algebraic (or analytical) ways of representing a function. Yet many of the examples of functions which could be so expressed had graphs which did not fit in with contemporary assumptions.

The representation of functions in the form of trigonometric series was introduced by Fourier in order to solve a problem in theoretical physics; but his results incidentally helped to undermine the naive way in which infinite processes were assumed to affect functions. For example, it is fairly easy to see that if we take two functions each of whose graphs forms a single continuous curve, such as $\sin x$ and $\frac{1}{3}\sin 3x$ (Figure 170), then their sum, $\sin x + \frac{1}{3}\sin 3x$, is also a function whose graph forms a single continuous curve. It follows that if we take any *finite* number of such functions each of whose graphs forms a single continuous curve, such as $\sin x$, $\frac{1}{3}\sin 3x$, $\frac{1}{5}\sin 5x$, ..., and $[\sin(2N-1)x/(2N-1)]$, then their sum

$$\sum_{n=1}^{N} \frac{\sin(2n-1)x}{2n-1}$$

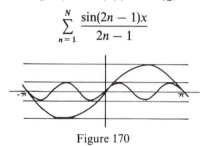

Figure 170

is also a function whose graph forms a single continuous curve. The naive view of *endless* sums, which fails to distinguish them from very long finite sums, takes it for granted that the same will be true if we take *endlessly* many functions, each of whose graphs forms a single continuous curve, such as $\sin x$, $\frac{1}{3}\sin 3x$, $\frac{1}{5}\sin 5x$, .... In blatant contradiction to this view we have the example

$$f(x) = \sum_{n=1}^{\infty} \frac{\sin(2n-1)x}{2n-1}$$

which is perfectly well defined for every value of $x$, but which has jumps at the points $x = k\pi$ (Figure 171). Each term in the endless sum has a graph which flows smoothly and continuously along; yet somehow forming their *endless* sum introduces totally unexpected "jumps."

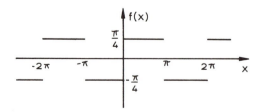

Figure 171

Such examples as this forced mathematicians to reconsider the *geometric* assumptions involved in their superficially *algebraic* image of a function. In particular they were forced to specify precisely what it meant to say that a function like $[\sin(2n - 1)x/(2n - 1)]$ was "*continuous*" (that is, that its graph is "all in one piece"), and to explain just when a combination of continuous functions will give rise to another continuous function, and when not. The formulation of *abstract* definitions, such as that of continuity forced mathematicians to reflect on the geometrical meaning of the ideas involved. The conventional wisdom, which saw a function as a more or less well-behaved algebraic expression, had confronted cold mathematical facts to produce a conceptual fog. Before it could find a path through this fog, "the science of *analytic* [or algebraic] expressions" was forced to remove its algebraic blinkers so as to take advantage of geometrical ideas about functions which could not be conveniently expressed in the conventional algebraic language. But though the geometrical image of a function was an important source of *psychological* inspiration in transcending the naive algebraic concept of a function, the *logical* presentation of the resulting ideas was if anything more formal and less geometrical than before.

Thus, for example, in the Introduction to his *Cours d'Analyse de l'École Royale Polytechnique*, based, as the title suggests, on the author's lectures at the École Polytechnique, Cauchy writes

> "I offer here the first part which goes under the title of *Algebraic Analysis*, and in which I treat in turn various kinds of functions, both real and imaginary ..."

The remaining parts were apparently never written. But however incomplete the work may have been, it marked a watershed in the struggle to bring some logical order into the use of infinite processes in analysis. The ideas presented were not all Cauchy's own, but he selected certain fundamental and carefully formulated ideas (such as the notions of *limit, convergence* and *continuity in an interval*), and used these to construct the framework for analysis on the basis of rigorous deduction alone. Implicit in the attempt itself was the view that many of the commonly accepted methods of deduction were in fact unreliable. In the Introduction we read

> "As for methods, I have sought to give them all the rigour that one requires in geometry, but in such a way as never to appeal to mere algebraic generalities. Appeals of this kind [...] attribute indefinite scope to algebraic formulae whereas in reality the majority of such formulae hold only under certain conditions and for certain values of the quantities involved."

Cauchy was objecting to the customary practice of manipulating algebraic expressions (such as $1/(1 - x) = 1 + x + x^2 + x^3 + ...$) without regard for the necessary restrictions on the value of $x$—which practice was often justified on the grounds that algebraic manipulations were *by their very nature universally valid.* Cauchy also wished to challenge the common assumption that, whenever an algebraic identity or formula was true for *real* number variables, one could simply assume the right to substitute *complex* variables if the need arose.

In the three hundred and thirty one pages of the *Cours d'Analyse* there are no diagrams and few appeals to geometry. One such appeal is implicit in the fact that the trigonometric functions are assumed to be defined in the classical way—that is, as lengths of certain lines associated with the unit circle (Figure 172). No such diagram appears in the *Cours d'Analyse*; but the number

$$\lim_{\alpha \to 0} \frac{\sin \alpha}{\alpha},$$

which one needs to know in order to differentiate the trigonometric functions, is evaluated by appealing to "obvious" *geometrical* facts (Exercise 6). However such "lapses" from formal algebraic argument are few and far between: indeed, in the one hundred and twenty pages of Chapters VII, VIII, and IX, where Cauchy develops the fundamental properties of complex numbers, complex variables and functions, and complex sequences, there seems to be not the slightest hint that complex numbers can be identified with points in the plane!

But though the official language of the *Cours d'Analyse* forbids open acknowledgement of the fact, the influence of geometrical ideas about functions, about continuity, and even about complex numbers is clearly visible on almost every page. Geometry is one of mathematics' richest sources of ideas, of intuitions, of analogies, of examples, of suggestive language, and of terminology; but ideas, intuitions, analogies, examples, and suggestive language do not fit comfortably into a coldly rigorous mathematical exposition,[17] and the unsuspecting reader may not notice their pervasive influence.[18]

The modern abstract function concept was not formulated once and for all by some clear sighted giant. We have seen that at least the germ of the idea appears to us to be visible in the words of Cauchy and Lagrange which we quoted earlier. The idea is even clearer in the 2nd edition of the influ-

---

[17] It is ironical that traditional school geometry was defended as being an ideal medium for teaching ideas of rigour and of proof. The singular virtue of geometry is that it *allows students to think*. The joy of traditional geometry was *solving the problem*—probably without writing anything beyond a sketch diagram; the ritual of writing out a proof added nothing, and usually subtracted a great deal.

[18] This tension between the coldly rigorous algebraic language of mathematical exposition and the underlying *meaning*, which is often best grasped unrigorously and geometrically, is even more acute today: nineteenth century mathematicians may have chosen to *write* algebraically, but they had at least been brought up to *think* geometrically.

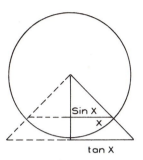

Figure 172

ential text *Traité du calcul differentiel et integral* (1810) by Lacroix (1765–1843). For there we read

"The classical analysts understood the expression *functions* of a quantity to refer to all the powers of this quantity. The meaning of this word was later extended so as to apply to the result of any number of algebraic operations; thus a function of one or several quantities was redefined to mean any algebraic expression made up in any manner from sums, products, quotients, powers and roots of these quantities. More recently, certain new ideas made possible by the progress of analysis have given rise to the following definition of functions.

*Any quantity whose value depends on one or on several other quantities, is called a function of these [other quantities], whether one knows or remains ignorant of the operations by which one passes from these [other quantities] to the first.*

The root of a quintic equation, for example, which in the present state of algebra cannot be given as an expression is nevertheless a function of the coefficients of the equation, because its value depends on the values of the coefficients ... It is not necessary that one has an equation between several quantities for one to say that one of them is implicitly a function of others. It is enough to know that its value depends on their particular values; thus in a circle, the *sine* is implicitly a function of the arc length, although algebraic analysis offers no means of expressing the relation between these two quantities, because the one is determined when the other is given and conversely."

Further on we find the following even more striking formulation:

"In order that a function of a single variable be completely determined, it is necessary and sufficient that for each particular value given to the variable one can deduce the corresponding value of the function."

One could scarcely hope for a clearer statement of the modern function-concept. Yet it would appear that in the ensuing sixteen hundred or so pages of Lacroix's three volume text the word "function" is consistently interpreted in the sense of "an algebraic formula or expression." Fourier also seems to have had the idea of an "arbitrary function": in his book

*Théorie analytique de la chaleur* (1822) we read

> "In general the function $f(x)$ represents a succession of values or ordinates each of which is arbitrary. An infinity of values being given to the abscissa $x$, there is an equal number of ordinates $f(x)$. All have actual numerical values, either positive, negative, or zero.
>
> We do not suppose that these ordinates be subject to a common law; they succeed each other in any manner whatever, and each of them is given as if it were a single quantity."

But though the examples discussed by Fourier are more general than those discussed by Lacroix, Fourier's examples in no way reflect the generality which we might be tempted to read into his 1822 definition. However, Dirichlet's investigations into the exact relation between a function $\varphi(x)$ and its Fourier series expansion

$$\tfrac{1}{2}a_0 + \sum_{n=1}^{\infty} (a_n \cos nx + b_n \sin nx),$$

which were published in 1829, were definitely based on a radically extended function-concept. In order to explain Dirichlet's result we must first make one or two elementary observations about Fourier series.

Given a function $\varphi(x)$ whose value is specified for each $x$ between $-\pi$ and $+\pi$, one can calculate the numbers $a_n$, $b_n$, where

$$a_n = \frac{1}{\pi} \int_{-\pi}^{+\pi} \varphi(x) \cos nx \, dx,$$

$$b_n = \frac{1}{\pi} \int_{-\pi}^{+\pi} \varphi(x) \sin nx \, dx.^{[19]}$$

These numbers $a_n$, $b_n$ can then be used to specify *the Fourier series of* $\varphi(x)$:

$$\tfrac{1}{2}a_0 + \sum_{n=1}^{\infty} (a_n \cos nx + b_n \sin nx).$$

A carefree application of the calculus leads one to the conclusion that *the value of the original function $\varphi(x)$ is equal to the value of its Fourier series for each $x$ between $-\pi$ and $+\pi$*. The steps which apparently lead to this conclusion were presented in Exercise 2(i)–(iii) of Chapter I.1 in the special case where

$$\varphi(x) = f(x) = \begin{cases} -1 & \text{if} \quad -\pi < x < 0 \\ \phantom{-}0 & \text{if} \qquad x = 0 \\ +1 & \text{if} \qquad 0 < x < \pi \end{cases}$$

---

[19] It was only later that mathematicians appreciated that this statement contains a crucial hidden assumption: namely the assumption that $\varphi(x)\cos nx$ and $\varphi(x)\sin nx$ *can in fact be integrated between $-\pi$ and $+\pi$.*

and gave rise to the Fourier series

$$\frac{4}{\pi} \sum_{n=1}^{\infty} \frac{1}{2n-1} \sin(2n-1)x.$$

Now in this particular case, *the function $f(x)$ is in fact equal to its Fourier series expansion for every value of $x$ between $-\pi$ and $+\pi$*; in particular $f(0)$ is equal to the value of the above Fourier series when $x = 0$. But suppose we modify the function $f(x)$ ever so slightly, by redefining its value at $x = 0$ to be equal to $\alpha$ say (where $\alpha \neq 0$), thereby obtaining a new function $f_\alpha(x)$:

$$f_\alpha(x) = \begin{cases} -1 & \text{if} & -\pi < x < 0 \\ \alpha & \text{if} & x = 0 \\ +1 & \text{if} & 0 < x < \pi \end{cases}$$

Then, since

$$a_n = \frac{1}{\pi} \int_{-\pi}^{+\pi} f(x)\cos nx \, dx = \frac{1}{\pi} \int_{-\pi}^{+\pi} f_\alpha(x) \cos nx \, dx,$$

and

$$b_n = \frac{1}{\pi} \int_{-\pi}^{+\pi} f(x)\sin nx \, dx = \frac{1}{\pi} \int_{-\pi}^{+\pi} f_\alpha(x)\sin nx \, dx,$$

$f_\alpha(x)$ has exactly the same Fourier series as $f(x)$. But then, since $f_\alpha(0) = \alpha \neq 0$, we see that *the value of $f_\alpha(x)$ is not equal to the value of its Fourier series expansion at the point $x = 0$.*

You may be inclined to suggest that, since $x = 0$ is a point of discontinuity of the graphs of both $f(x)$ and $f_\alpha(x)$, it is bound to be exceptional. But, as we have pointed out, discontinuities only appear exceptional if one thinks in terms of a naive function-concept, whether it be naively geometric ("all functions have graphs which can be drawn by hand") or naively algebraic ("all functions behave like the familiar elementary functions").

Fourier series had demonstrated their importance both in the solution of interesting problems in applied mathematics and in the resolution of fundamental questions about functions. It was therefore important to investigate the precise relationship between the value of a function $\varphi(x)$ at the point $x$ and the value of its Fourier series expansion. But before we can state the result of Dirichlet's investigations into this relationship between $\varphi(x)$ and its Fourier series, we must introduce an important, but intuitively simple, idea.

If the point $x = c$ is a jump-point for the graph of the function $\varphi(x)$, then the value $\varphi(c)$ of the function at the point $x = c$ tells us absolutely nothing about the behaviour of the function $\varphi(x)$ *near* the point $x = c$. Much more informative in this respect are the two values $\varphi(c+)$ and $\varphi(c-)$ whose meaning we will now explain. The first of these is *the value which one might expect* the function $\varphi(x)$ to take at the point $x = c$ if one looked only at values $x > c$, and asked

*Where is $\varphi(x)$ heading for as $x$ approaches $c$ from above?*

For example, when $\varphi(x)$ is the function $f(x)$ (or $f_a(x)$) and $c = 0$ (Figure 173), then as $x$ approaches $c = 0$ *from above*, $f(x)$ remains constant with the value 1: in other words, $f(x)$ *is heading for the value 1 as x approaches 0 from above*—a fact which we express by the equation

$$f(0+) = 1.$$

Similarly, as $x$ approaches 0 *from below*, $f(x)$ remains constant with the value $-1$; in other words, $f(x)$ *is heading for the value $-1$ as x approaches 0 from below*—a fact which we express by the equation

$$f(0-) = -1.$$

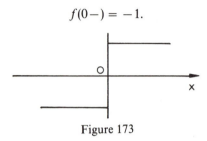

Figure 173

The fact that $f(0+)$ and $f(0-)$ are different in this case expresses the clearly visible fact that the point $x = 0$ is a jump-point in the graph of the function $f(x)$. In general we will have $\varphi(c+) = \varphi(c-)$ unless $x = c$ happens to be a jump-point in the graph of the function $\varphi(x)$.

We can now state Dirichlet's result:

"If the function $\varphi(x)$, whose values are assumed to be all finite and determined, has only finitely many discontinuities between $-\pi$ and $+\pi$, and if there are only a finite number of maxima and minima between these bounds, then the Fourier series of $\varphi(x)$ has the value

$$\tfrac{1}{2}\bigl(\varphi(x+) + \varphi(x-)\bigr)$$

at each point $x$ between $-\pi$ and $+\pi$."

Now this is in many ways a very satisfactory result; but Dirichlet realised that however reasonable its hypotheses (that $\varphi(x)$ should have only finitely many discontinuities and only finitely many maxima and minima between $-\pi$ and $+\pi$) may appear from the point of view of a naive function-concept, they were in fact very restrictive. To illustrate the extent to which a function might fail to satisfy these conditions he gave a radically simple example of a function whose values were all finite and determined: namely, for $x$ between $-\pi$ and $+\pi$ he defined

$$\varphi_1(x) = \begin{cases} c & \text{if } x \text{ is rational} \\ d & \text{if } x \text{ is irrational} \end{cases}$$

where $c$ and $d$ may be any two distinct real numbers. This example gives the unmistakeable impression that Dirichlet had well and truly transcended the idea that a function is necessarily given (either explicitly or implicitly) by some algebraic formula. This impression is further supported by Dirichlet's criticism, in the same 1829 paper, of Cauchy's own analysis of the problem. For Cauchy had assumed that it was permissible to substitute complex values for $x$ in the given function $\varphi(x)$, whereas Dirichlet observes that such an assumption only makes sense if $\varphi(x)$ is given by some algebraic formula. But however general Dirichlet's own function-concept may have been, he can scarcely have imagined the incredible variety of functions which would eventually have to be taken into account.[20]

To end this chapter let us try to summarise what we earlier labelled the "creative tug-of-war" between the geometric and the algebraic viewpoints. We observed that the calculus had its roots in the study of geometric curves, and that it was the introduction of coordinates which allowed these problems about curves to be expressed in the algebraic language of "variables" $x$ and $y$ constrained to satisfy certain equations. This formulation of superficially distinct geometric problems in a uniform algebraic way encouraged the development of uniform methods of solution—culminating eventually in the remarkably effective, but logically inconsistent, calculi of Newton and Leibniz. At first the problems which were tackled by means of the calculus tended to concern variables with strong geometrical or mechanical meanings; but as mathematicians' manipulative facility increased, and as the problems they studied became more complex, the calculus became essentially a way of operating on algebraic variables and expressions. And when the calculus eventually came to be seen as a calculus of functions, the function-concept which accompanied this change of viewpoint was formally algebraic, in that it assumed the function to be specified by some algebraic formula; but at the same time it was implicitly assumed that all such functions shared the pleasant geometrical properties of the familiar elementary functions and the classical geometric curves. This naive

---

[20] This is indicated, for example, by the fact that Dirichlet does not appear to have observed that in the case of his "rational/irrational" counter-example $\varphi_1(x)$ one cannot even calculate the Fourier coefficients $a_n$, $b_n$. The same sort of conclusion may be inferred from Dirichlet's comments in a letter to Gauss in 1853, where he writes: "I find that your conjecture [that a function with infinitely many maxima and minima between $-\pi$ and $+\pi$ might also have a Fourier series expansion] is fully justified provided that one excludes certain very curious cases." Dirichlet gives an abstract characterisation of these cases without offering any examples. (This letter is referred to in the book *Proofs and Refutations* by I. Lakatos, p. 150, during a discussion in which the author makes a number of interesting points: in particular he points out that Dirichlet never in fact published the modern definition of a function. I have not checked this assertion thoroughly, but it is clear that Morris Kline is wrong when he claims (*Mathematical Thought from Ancient to Modern Times*, Oxford University Press, (1972), p. 950) that, "Dirichlet gave the definition of a ... function which is now most often employed [in his 1837 paper entitled 'On the representation of completely arbitrary functions by sine and cosine series']". Perhaps because this paper was written for physicists, Dirichlet there restricts attention to *continuous* functions; but he makes no attempt to define "continuous", preferring instead to appeal to his readers' intuitive idea of a *geometric curve*. The "completely arbitrary functions" of the title are in fact "arbitrary geometric curves." However it seems to me that Lakatos goes too far, for example, when he asserts that "there is ample evidence that [Dirichlet] had no idea of [the modern function] concept.")

algebraic image of a function was gradually undermined—its eventual refinement being at least partly the result of a return to the more flexible, if
less precise, geometric image of a function. But though the geometric viewpoint may have been psychologically important in showing how the calculus might be given a solid foundation, it was eventually excluded from the
logical structure of what we have referred to as "the 1870 version of the
calculus."

But the geometric viewpoint could not be kept out for long. The 1870
version of the calculus was soon found to have important technical defi-
ciencies, stemming mainly from the definition and properties of the Riemann integral. The sequence of events which finally got round these defi-
ciencies (leading up to Lebesgue's theory of integration) represents an even
more striking example of the contribution which the geometric viewpoint
has made to the evolution of modern analysis. The interested reader will
find these events beautifully analysed and presented in the book *Lebesgue's
Theory of Integration* by Thomas Hawkins (Chelsea Publishing Co. 1975).

EXERCISES

1. It is sometimes possible to invent an equation for an unfamiliar curve provided
   one is prepared to introduce new functions. For example, suppose for each real
   number $x$ we define

   $$[x] = \text{"the largest whole number } \leq x\text{"}$$

   and

   $$\{x\} = x - [x].$$

   (i) Sketch the graphs of

   $$y = [x], \quad y = \{x\}, \quad y = \{x\} - \tfrac{1}{2},$$
   $$y = |\{x\} - \tfrac{1}{2}|, \quad y = [2\{x\}].$$

   (ii) Sketch the graph of the endless sum

   $$y = \sum_{n=1}^{\infty} \frac{\sin(2n - 1)x}{(2n - 1)}$$

   which we first met in Chapter I.1. Use the functions $[x]$ and $\{x\}$ to obtain a
   "finite" expression which is equal to this endless sum for all values of $x$ except
   integer multiples of $\pi$.

   (iii) Suppose that

   $$y = \tfrac{1}{2}a_0 + \sum_{n=1}^{\infty} a_n \cos nx$$

   has a graph which looks like that in Figure 174. Use a method similar to
   that described in Exercise 2(iii) of Chapter I.1 to find the coefficients $a_0, a_1, a_2,$
   $a_3, \dots$ . Then use the functions $|x|$ and $\{x\}$ to obtain a "finite" expression which
   is equal to this endless sum for all values of $x$. Finally use the values of $a_0, a_1, a_2,$
   $a_3, \dots$ which you obtained, together with the fact that $y = 0$ when $x = 0$, to

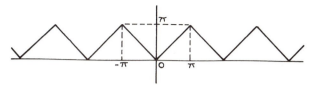

Figure 174

obtain the formula

$$\frac{\pi^2}{8} = 1 + \frac{1}{3^2} + \frac{1}{5^2} + \frac{1}{7^2} + \dots .$$

(iv)  Can you find simple expressions for the functions whose graphs are shown in Figure 175?

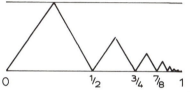 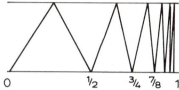

Figure 175

2.  Put $f_0(x) = |x|$ when $-\pi \le x < \pi$; for values of $x$ outside this range simply write $x$ in the form $x = x' + 2k\pi$, where $-\pi \le x' < \pi$, and define $f_0(x) = f_0(x')$. Thus $f_0(x)$ has the graph shown in Figure 174. For $n \ge 1$, put $f_n(x) = f_0(4^n x)/4^n$.
(i)  Sketch the graphs of

$$y = f_0(x), \quad y = f_1(x), \quad y = f_2(x)$$

on the same pair of axes. Show that $f_n(x)$ has maximum value $\pi/4^n$ and that its graph simply repeats itself periodically with period $2\pi/4^n$.
(ii)  Define the function $f(x)$ by

$$f(x) = \sum_{n=0}^{\infty} f_n(x) = \sum_{n=0}^{\infty} f_0(4^n x)/4^n$$

and find the values of $f(0), f(\pi/2), f(\pi)$.
(iii)  Sketch the graphs of

$$y = f_0(x), \quad y = f_1(x), \quad y = f_0(x) + f_1(x)$$

for $0 \le x \le 2\pi$ on one pair of axes. Sketch the graphs of

$$y = f_0(x) + f_1(x), \quad y = f_2(x), \quad y = f_0(x) + f_1(x) + f_2(x)$$

for $0 \le x \le 2\pi$ on one pair of axes.
(iv)  Let $\varepsilon > 0$. Show that

$$|f_n(x \pm \varepsilon) - f_n(x)| < \varepsilon$$

for each $n = 0, 1, 2, \ldots$. Use this to show that

$$|f(x \pm \varepsilon) - f(x)| < 4\varepsilon/3$$

and hence conclude that $f(x)$ is *continuous at each point* $x$.

Next we show that $f(x)$ is *not differentiable at any point* $x$. To say that "$f(x)$ is differentiable at $x$" means simply that

$$\text{"}\lim_{h \to 0} \frac{f(x + h) - f(x)}{h} \quad \text{exists"}.$$

(v) Let the point $x$ be given. Choose any $n \geq 0$; then $h_n$ is chosen to be either $+\pi/4^{n+1}$ or $-\pi/4^{n+1}$ so that the points $(x, f_n(x))$ and $(x + h_n, f_n(x + h_n))$ both lie on the same straight segment in the graph of $y = f_n(x)$. Show that

$$|f_m(x + h_n) - f_m(x)| = |h_n| \quad \text{for every } m \leq n$$

and

$$|f_m(x + h_n) - f_m(x)| = 0 \quad \text{for every } m > n.$$

Hence show that $(f(x + h_n) - f(x))/h_n$ is a non-zero integer, and that this integer is odd when $n$ is even, and even when $n$ is odd. Conclude that the function $f(x)$ is not differentiable at any point $x$.

3. Cauchy had defined the definite integral of a *continuous* function. The same idea allows one to integrate discontinuous functions provided the jump-points are well spaced out by simply adding up the integrals on the separate continuous pieces. Riemann (1826–1866) introduced a much more general definition of the integral and gave the following example in order to exhibit a function which was integrable in his sense, but not in Cauchy's. However, we shall not discuss this aspect of Riemann's example. What interests us here is the fact that Riemann's example is of a function $f(x)$ whose graph makes a jump of height $\pi^2/8q^2$ at each point $x = p/2q$, where $p$ and $q$ are integers having no common divisors, and $p$ is odd.

(i) Define the function $\phi(x)$ by

$$\phi(x) = \begin{cases} x - N & \text{when } N \text{ is the } \textit{unique} \text{ integer closest to } x \\ 0 & \text{when } N = \pm\tfrac{1}{2}, \pm\tfrac{3}{2}, \tfrac{5}{2}, \ldots \end{cases}$$

Sketch the graph of $\phi(x)$.

(ii) Show that $\phi(x) = \{x + \tfrac{1}{2}\} - \tfrac{1}{2}$ provided $x \neq \pm\tfrac{1}{2}, \pm\tfrac{3}{2}, \pm\tfrac{5}{2}, \ldots$. (The meaning of $\{x + \tfrac{1}{2}\}$ is explained in Exercise 1.)

(iii) *Assume* that $\phi(x)$ can be expanded as an endless sum of sines and use the method described in Exercise 2(iii) of Chapter I.1 to find the coefficients $b_n$.

(iv) Define the function $\phi_n(x)$ by

$$\phi_n(x) = \phi(nx).$$

Sketch the graph of $y = \phi_n(x)$, and show that the graph has jumps at $\pm 1/2n$, $\pm 3/2n, \pm 5/2n, \ldots$.

(v) Using the definition of $\phi(c+)$ and $\phi(c-)$ on page 273, show that, provided $c \neq \pm 1/2n, \pm 3/2n, \pm 5/2n, \ldots$, we have

$$\phi_n(c+) = \phi_n(c-) = \phi_n(c)$$

(in other words, $\phi_n$ is continuous at the point $c$).

(vi) Now define the function $f(x)$ to be the following endless sum:

$$f(x) = \phi_1(x) + \frac{\phi_2(x)}{2^2} + \frac{\phi_3(x)}{3^2} + \ldots + \frac{\phi_n(x)}{n^2} + \ldots .$$

*For this particular endless sum one can in fact prove that*

$$f(x+) = \left( \sum_{n=1}^{\infty} \frac{\phi_n}{n^2} \right)(x+) = \sum_{n=1}^{\infty} \frac{\phi_n(x+)}{n^2},$$

and

$$f(x-) = \left( \sum_{n=1}^{\infty} \frac{\phi_n}{n^2} \right)(x-) = \sum_{n=1}^{\infty} \frac{\phi_n(x-)}{n^2}.$$

Use this to show that, if $x = p/2q$, where $p$ and $q$ are integers having no common factors and $p$ is odd, then

$$f(x+) - f(x) = -\frac{1}{2q^2}\left( 1 + \frac{1}{3^2} + \frac{1}{5^2} + \ldots \right) = -\frac{\pi^2}{16q^2}$$

and

$$f(x-) - f(x) = +\frac{1}{2q^2}\left( 1 + \frac{1}{3^2} + \frac{1}{5^2} + \ldots \right) = +\frac{\pi^2}{16q^2}$$

whereas, if $x$ is not of this form, then

$$f(x+) = f(x-) = f(x).$$

4. Suppose that $f(x)$ is defined for all real numbers $x \neq 0$ and that, for each $x > 0$, it satisfies two conditions

$$f(x) = \ln x$$

$$\frac{d}{dx} f(-x) = \frac{d}{dx} f(x).$$

Convince yourself that $f$ is uniquely determined once $f(-1)$ is given, and that for each $x > 0$

$$f(-x) = -f(x) + f(-1).$$

5. Euler eventually resolved the dispute about the value of $\log(-1)$ by showing what no-one had suspected: namely, that whatever value of $x$ one chooses (other than $x = 0$), there are *infinitely many* values of $\log x$; when $x > 0$, just one of these values is real (hence the graph), whereas all the other values are complex; and when $x < 0$ none of the values of $\log x$ is real (hence the gaping hole). Euler's argument is a typically daring calculation involving both infinitely large and infinitely small quantities: we give it here in outline as another example of the kind of double-edged success which made mathematicians so reluctant to reject vague arguments involving infinitely large and infinitely small quantities. Like the methods of naive calculus, Euler's argument is strictly indefensible, yet it delivers the goods in a quite remarkable way. You are invited to fill in the details.
     Let $y = \log x$ and put $x = (1 + \omega)^n$ where $n$ is infinitely large and $\omega$ is infinitely small. Then $y = n \cdot \omega$. But $x = (1 + \omega)^n$ implies $n\omega = n(\sqrt[n]{x} - 1)$, so

$\log x = n(\sqrt[n]{x} - 1)$ (where $n$ is infinitely large). Just as $\sqrt[2]{x}$ has two values, and $\sqrt[3]{x}$ has three values, so $\sqrt[n]{x}$ will have infinitely many values. Hence $\log x$ *also has infinitely many values*. Suppose, for example, that $x = 1$. Then $y = \log 1 = n(\sqrt[n]{1} - 1)$, so $(1 + (y/n))^n - 1 = 0$. Thus to find the $n$ values of $y$ we must factorise $(1 + (y/n))^n - 1$ into $n$ factors. But combining conjugate roots of $z^n - 1$ gives a real quadratic factor of the form $z^2 - 2z \cos(2\pi\lambda/n) + 1$ for each $\lambda \in \mathbb{Z}$, corresponding to the linear factors $z - (\cos(2\pi\lambda/n) \pm \sqrt{-1} \sin(2\pi\lambda/n))$. Thus putting $z = 1 + (y/n)$ (and remembering that $n = \infty$, so that $\cos(2\pi\lambda/n) = 1$, $\sin(2\pi\lambda/n) = (2\pi\lambda/n)$) we get the infinitely many values of $\log_e 1 = y = 2\pi\lambda\sqrt{-1}$, one for each integer value of $\lambda$.

6. Use the classical definitions of $\sin x$, $\cos x$, $\tan x$ (for $0 \le x < \pi/2$) implicit in Figure 172 to construct a plausible argument justifying each of the following steps:

(a) $\tan x > x > \sin x$.

(b) $1 > \dfrac{\sin x}{x} > \cos x$.

(c) $\displaystyle\lim_{x \to 0} \dfrac{\sin x}{x} = 1$.

7. Some insight into the way one is inevitably forced to extend a *naively algebraic function-concept* may be gained by reflecting on our own experience of elementary functions and the operations we perform on them. We shall consider only the operations of *adding* (two or more) functions, *multiplying* functions, forming the *quotient* of two functions, *composing* two functions, taking the *inverse* of a function, and *differentiating* and *integrating* a given function. In particular, we shall not go so far as to consider the two operations which were so crucial in constructing the examples in Exercises 2 and 3: that is, we shall consider neither *splicing separate algebraic pieces together*, nor *forming endless sums*.

   We shall start with the most basic functions we can think of: that is, the *identity function i*, defined by

$$i(x) = x \quad \text{for every } x$$

and the *constant functions* **c** (where $c$ may be any real number), defined by

$$c(x) = c \quad \text{for every } x.$$

For example, the constant function **2** takes the value $\mathbf{2}(x) = 2$ at every point $x$ (so its graph is the line $y = 2$).
   In the text we have been rather sloppy in referring to

$$\text{``the function } \tfrac{1}{3} \sin 3x \text{''} \quad \text{and} \quad \text{``the function } f(x)\text{''}.$$

Strictly speaking, the functions here are

$$\text{``} \tfrac{1}{3} \sin 3(-) \text{''} \quad \text{and} \quad \text{``} f(-) \text{''} \quad (\text{or ``} f \text{'' for short}),$$

whereas "$\frac{1}{3}$ sin $3x$" and "$f(x)$" are the *values* of these functions at the point $x$.[21] If $f$ and $g$ are functions, then

their *sum* $f + g$ is the function whose value $(f + g)(x)$ at $x$ is $f(x) + g(x)$;
their *product* $f * g$ is the function whose value $(f * g)(x)$ at $x$ is $f(x) \cdot g(x)$;
their *quotient* $f/g$ is the function whose value $(f/g)(x)$ at $x$ is $f(x)/g(x)$;

and

their *composition* $f_o\, g$ is the function whose value $(f_o\, g)(x)$ at $x$ is $f(g(x))$.

(i) (a) Show that $i + i = 2 * i$ and $i + (-1 * i) = 0$.
    (b) Work out $(2 * i + -3)(x)$ and $(i * i + 3 * i + -4)(x)$.

To simplify notation, write

$$i^2 = i * i, \quad \text{and} \quad i^{n+1} = i^n * i \quad \text{when } n \geq 2.$$

(ii) Let $f$ be the *polynomial* function with coefficients $a_0, a_1, \ldots, a_n$:

$$f(x) = a_n x^n + a_{n-1} x^{n-1} + \ldots + a_1 x + a_0.$$

Write down an expression (involving powers of the identity function $i$ and suitable constant functions) which is equal to $f$. Thus if we start with the identity function $i$ and the constant functions **c**, we can obtain *all polynomial functions* by using just the operations of addition $(+)$ and multiplication $(*)$.

(iii) What kinds of functions does one get by adding and multiplying polynomial functions?

What kinds of functions does one get by differentiating and integrating polynomial functions?

(iv) (a) Show that

$$(i * i + 3 * i + -4)_o(2 * i + -3)(x) = (2x - 3)^2 + 3 \cdot (2x - 3) - 4.$$

    (b) Write the function $(i * i + 3 * i + -4)_o(i * i + 3 * i + -4)$ as a polynomial function.

    (c) Explain why the composition $f_o\, g$ of two polynomial functions $f$ and $g$ is always a polynomial function.

The quotient of two polynomial functions will not usually be a polynomial function: for example

$$h = \frac{1}{i} \quad \left(h(x) = \frac{1}{x}\right), \quad \text{and} \quad k = \frac{1}{i^2 + 1} \quad \left(k(x) = \frac{1}{x^2 + 1}\right)$$

are not polynomial functions. In general, the quotient $f/g$ of two polynomial functions $f$ and $g$ is called a *rational function* (provided that $g$ is not the zero polynomial **0**). Thus a function $h$ is a rational function provided there exist two fixed polynomials $f$ and $g$ (with $g \neq \mathbf{0}$) such that

$$h(x) = \frac{f(x)}{g(x)} \quad \text{for every } x.$$

(v) Show that if $h = f/g$ and $k = p/q$ are two rational functions, then their sum

---

[21] Even this is a half-truth: see, for example, "Clarity in the calculus: or Leibniz ab naevo vindicatus" by Hugh Thurston in *Mathematics Teaching* **67** (1974).

$h + k$, their product $h * k$, and their quotient $h/k$ are all rational functions too. Is their composition $h_0 k$ a rational function?

(vi) What kind of function does one get when one differentiates a rational function $h = f/g$?

Though polynomial functions could always be integrated to give other polynomial functions, the same is not always true for rational functions. In most cases, the attempt to integrate rational functions forces us to introduce *new* functions: for example

$$\int \frac{1}{x}\, dx = \ln(x) \quad \text{and} \quad \int \frac{1}{x^2 + 1}\, dx = \text{arc tan } x.$$

*It is a remarkable, and essentially elementary, result that if we allow algebraic expressions involving arbitrary rational functions and just these two new functions "ln" and "arc tan," then it becomes possible to write down the integral of any rational function.*

An excellent treatment of this result may be found in the book *Differential and Integral Calculus*, Vol. 1, by Richard Courant (Chapter IV, Section 5, entitled "Integration of Rational Functions"). The central idea is simply to rewrite the expression $f(x)/g(x)$ as the sum of a polynomial $p(x)$ and so-called *partial fractions* of the form

$$\frac{A}{(x - a)^m} \quad \text{and} \quad \frac{B + Cx}{((x + b)^2 + d^2)^n}$$

(vii) Show that the integrals of all such partial fractions can be reduced to integrals of the form

$$\int \frac{1}{y^m}\, dy \quad \text{and} \quad \int \frac{1}{(y^2 + 1)^n}\, dy.$$

Evaluate these integrals. [Hint: For the second, either substitute $y = \tan t$ and derive a recurrence relation, or use

$$\frac{1}{(y^2 + 1)^n} = \frac{1}{(y^2 + 1)^{n-1}} - \frac{y^2}{(y^2 + 1)^n}$$

to obtain a recurrence relation straight off.]

We have now been forced to accept not only rational functions, but also the two new functions "ln" and "arc tan." By forming sums, products, quotients and compositions of these new functions with rational functions we obtain an even larger family of functions. It is slightly surprising that *some* of these combinations can be integrated without introducing any new functions.

(viii) Evaluate the following integrals:

$$\int \text{arc tan } x\, dx, \qquad \int \ln(x)\, dx, \qquad \int \frac{1}{x} \ln(x)\, dx,$$

$$\int x^a \ln(x)\, dx \quad (a \neq -1).$$

However this is not true in general—even for such harmless-looking examples as the so-called *logarithmic integral* $\int (1/\ln(x))\ dx$.[22]

The appearance of the functions "ln" and "arc tan" serves to remind us that we have not yet explored the operation of *taking inverses*: in particular "*arc* tan" has appeared *before* "tan" (or "sin" or "cos"), and "ln" has appeared *before* the exponential function. Suppose we now go right back to the beginning and see what happens when we start taking inverses.

If we consider the original functions "*i*" and "**c**" then we see that *i* is its own inverse, whereas **c** does not have an inverse. Simple linear functions have similar looking inverses (e.g. $2 * i + 3$ has inverse $(\frac{1}{2}) * (i + -3)$). But if we want a quadratic function like $i^2$ $(i^2(x) = x^2)$ to have an inverse, then we are already forced to introduce a new function—which we shall call "$i^{1/2}$", and which is defined by

$$i^{1/2}(x) = +\sqrt{x}\quad \text{(for every } x \geq 0\text{)}.$$

Observe that whereas in the early stages of this Exercise we got away with the pretence that every function "*f*" was defined for (more or less) every value of *x*, and even failed to observe that "ln" is only defined for $x > 0$, the introduction of inverse functions forces one eventually to become sensitive to the values of *x* for which the function is defined. Moreover, though a simple polynomial *f* automatically produces a single value $f(x)$, we are now faced with a choice: "Should $i^{1/2}(x)$ denote the positive square root, the negative square root, or both?"

(ix) Find an inverse of $3 * i^2 + 2 * i + 1$. For which values of *x* is this inverse defined?

We may clearly go on to consider cube roots, $n^{\text{th}}$ roots, and inverses of rational functions in general, but in practice, even the simplest combination of square roots with polynomial or rational functions breaks new ground.

(x) Evaluate the following integrals

$$\int \sqrt{a^2 - x^2}\ dx, \qquad \int \frac{1}{\sqrt{a^2 - x^2}}\ dx, \qquad \int \sqrt{a^2 + x^2}\ dx.$$

In reality, these and other standard integrals which one learns to evaluate in a calculus course give a misleading impression. Thus in Courant's book we read

"Attempts to express general integrals such as

$$\int \frac{dx}{\sqrt{(a_0 + a_1 x + \ldots + a_n x^n)}},$$

$$\int \sqrt{(a_0 + a_1 x + \ldots + a_n x^n)}\ dx, \quad \text{or} \quad \int \frac{e^x}{x}\ dx$$

in terms of elementary functions have always ended in failure; and in the nineteenth century it was finally proved that it is actually impossible to carry out these integrations in terms of elementary functions.

---

[22] The logarithmic integral is equivalent to the example $\int (e^t/t)\ dt$ discussed on page 265; put $x = e^t$.

If, therefore, the object of the integral calculus were to integrate functions in terms of elementary functions, we should have come to a definite halt. But such a restricted object has no intrinsic justification; indeed it is of a somewhat artificial nature ... .

Where the integral of a function cannot be expressed by means of functions with which we are already acquainted, there is nothing to hinder us from introducing this integral as a new "higher" function of analysis, which really means no more than giving it a name."

(xi) Read the whole of Chapter IV, Section 7 in Courant: *Differential and Integral Calculus*, Vol. 1 (from which the above quotation is an extract).

8. The circle with equation

$$x^2 + y^2 = a^2$$

has area $\pi a^2$ and circumference $2\pi a$.

Show that the ellipse with equation

$$\frac{x^2}{a^2} + \frac{y^2}{b^2} = 1$$

has area $\pi ab$.

Show that the circumference of this ellipse is given by

$$4 \int_0^a \sqrt{\frac{a^4 - x^2(a^2 - b^2)}{a^2(a^2 - x^2)}} \, dx.$$

(This is an example of an integral which cannot be evaluated in terms of elementary functions.)

9. Calculate the first seven derivatives of $f(x) = \tan x$, expressing each as a polynomial in $\tan x$. Hence check the terms of the series for $\tan x$ given in the text. The coefficients of these derivatives can be tabulated as follows

|            | $\tan^0 x$ | $\tan x$ | $\tan^2 x$ | $\tan^3 x$ | $\tan^4 x$ | ... |
|------------|-----------|----------|-----------|-----------|-----------|-----|
| $f(x)$     | 0         | 1        | 0         | 0         | 0         | ... |
| $f^{(1)}(x)$ | 1       | 0        | 1         | 0         | 0         | ... |
| $f^{(2)}(x)$ | 0       | 2        | 0         | 2         | 0         | ... |
| $f^{(3)}(x)$ | 2       | 0        | 8         | 0         | 6         | ... |
| $f^{(4)}(x)$ | 0       |          |           |           |           |     |

It is easy to see why the terms 1, 1, 2, 6, ... down the "right hand side" are just 0!, 1!, 2!, 3!, ..., and that the terms in the $\tan^0 x$ and $\tan x$ columns pair off as shown by the arrows. Can you find a general rule which will allow you to determine all the coefficients of $f^{(1)}(x)$ once you know those of $f(x)$, all the coefficients of $f^{(2)}(x)$ once you know those of $f^{(1)}(x)$, etc. without differentiating? Use your rule to find the terms of the power series for $\tan x$ up to the term in $x^{19}$.

CHAPTER IV.3

# What Is an Exponential Function?

We shall end Part IV by examining very briefly one particular class of functions: namely *powers* $x^\alpha$, otherwise known as *exponential functions*. Our aim in so doing is simply to indicate the richness and the complexity of our own mathematical experience of such functions, and to consider how this complex experience might lead us eventually to appreciate the way exponential functions are usually treated in an analysis course.

It is not often that we have the chance as students to *reinvent* standard mathematical conventions. One notable exception occurs when, having met (and understood) zero, fractions, negative numbers, roots, ordinary powers, and those marvellous labour saving devices—the index laws

$$x^m \cdot x^n = x^{m+n}, \qquad (x^m)^n = x^{mn}$$

for positive integer exponents $m$, $n$, we have the opportunity to exercise our imagination in looking for possible interpretations of *negative* and *fractional* powers. Once we have seen how the index laws suggest sensible meanings for, say, $2^{-1}$, $4^{1/2}$ and $8^{2/3}$, there is a certain fascination in trying to make sense of such expressions as

$$32^{2/5}, \quad 16^{1\frac{1}{4}}, \quad 4^{3\frac{1}{4}}, \quad \left(\frac{125}{64}\right)^{-2/3}, \quad (2\tfrac{1}{4})^{-1/2}, \quad (-1)^{-1/3}.$$

We may even succeed in explicitly formulating the rules

$$x^{p/q} = \left(\sqrt[q]{x}\right)^p = \sqrt[q]{(x^p)},$$

or

$$\text{``}x^{p/q}\text{'' means} \begin{cases} \text{``the } p^{\text{th}} \text{ power of the } q^{\text{th}} \text{ root of } x\text{''} \\ \text{``the } q^{\text{th}} \text{ root of the } p^{\text{th}} \text{ power of } x\text{''}. \end{cases}$$

If we do, then it is exceedingly unlikely that anyone will encourage us to examine the curious nature of our discovery. For example, if we try to work out $(-1)^{2/6}$, we find that whereas $(\sqrt[6]{-1})^2$ cannot be evaluated, $\sqrt[6]{(-1)^2}$ is equal to $\pm 1$: and neither of these agrees with $(-1)^{1/3} = -1$. Consideration of such examples would eventually lead us to the conclusion that our rule seemed to give consistent results *provided that p and q are integers having no common factors, and that $q > 0$*. But we should at the same time come to realise that there was something unsatisfactory about a rule which gives rise

285

to a different answer (or even no answer at all) when the actual value of the exponent $1/3$ is not changed but simply written in the form $2/6$. In the absence of such observations as this we get so strongly, and unquestioningly, wedded to the idea that exponentials $x^\alpha$ can be interpreted in terms of integral powers and roots, that when we eventually meet the official definition

$$\text{``} x^\alpha = e^{\alpha \ln x} \quad (\text{provided } x > 0) \text{''}$$

it leaves hardly any impression. Indeed, it looks more like a rather boring tautology than a genuine *definition* of the elusive function $x^\alpha$: after all don't we know already that

$$\alpha \ln x = \ln x^\alpha$$

and that

$$e^{\ln y} = y?$$

What then is all the fuss about? Unimpressed we soldier on with our well-tried "exponential tool-kit" consisting of the familiar index laws ($x^\alpha \cdot x^\beta = x^{\alpha+\beta}$, $(x^\alpha)^\beta = x^{\alpha \cdot \beta}$—now presumed to work for *all* exponents $\alpha$ and $\beta$), and the familiar rule for evaluating $x^{p/q}$ as the "$q^{\text{th}}$ root of the $p^{\text{th}}$ power of $x$."

And this works remarkably well as long as we simply assume that we can evaluate $2^{\sqrt{2}}$ by approximating $\sqrt{2}$ by a fraction $p/q$, or by simply using the "$y^x$" button on a pocket calculator.

But once one begins to ask questions about the mathematical meaning of such expressions as $2^{\sqrt{2}}$, $2^\pi$ and even $(\sqrt{2})^\pi$, it becomes clear that there is something missing. Consider, for example, the following letter from the journal *Mathematics Teaching* **86** (1979):

> "Dear Sir,
>
>   "Most of us have taught some things about the beginning of exponential ideas by having the graph of $y = 2^x$ drawn. It goes like this: you know where the [positive] integral powers are [...]; if we retain the multiplication rule for indices we can easily see what the fractional and negative indices mean, so away we go and join them up in one fine continuous curve. Am I the only one who has had a little voice at the back of his mind saying 'If you believe that you'll believe anything'?
>   "There are two points in this sequence which I find a bit disturbing. I do try to come clean with a good class on the first, namely that even roots may be positive or negative so that there is a shadow of the curve below the x-axis. The other point is really a question I cannot answer: what are we to say about irrational [and even] transcendental indices? Of course I can find approximate numerical values for $2^{\sqrt{2}}$ and $2^\pi$ but in doing so I am assuming that they behave like ordinary indices, which they quite clearly are not. We attribute 'meaning' to negative and fractional indices; can someone (can anyone?) do as much for the others I have mentioned? And if not do we have to say that $y = 2^x$ is not, after all, a continuous function so that all talk of differentiating it is without foundation.

"In an effort to clear my mind about this I decided to invert and see what happened to the graph of $y = x^n$ as $n$ varied. I supposed that I would see a nice smooth transition from one curve to the next, the sort of thing that would make a very pretty film. I found nothing of the sort. I had in mind having a close look at the interval $n = 0$ to $n = 1$ in which, of course, I would see the line $y = 1$ growing steadily and interestingly into the line $y = x$. To my irritation and surprise I did not find it easy even to 'take off.' It seems to me that as $n$ tends to zero some odd things are about to happen which can be illustrated by the observation that for $n = 1/2$ the graph seems to be in the first and fourth quadrants, for $n = 1/3$ it seems to be in the first and third quadrants and for $n = 2/3$ it seems to be in the first and second quadrants [...]. Life becomes even more confusing if one takes heed of the [fact that $((-1)^2)^{1/6}$ and $((-1)^{1/6})^2$ are different]. We have all taught (have we not?) that $\sqrt[n]{a^m} = (\sqrt[n]{a})^m$. Well is it?

"It all seems very odd to me and if someone is able to organise things in a systematic way please let it be done; but to return to my first question, what *are* we to say (or even to think) about $2^{\sqrt{2}}$ and $2^\pi$?"

The present chapter may be viewed as an attempt to respond to this challenge "*to organise* [*exponential functions*] *in a systematic way.*" It is definitely *not* a *logically* systematic treatment of exponentials: there is no shortage of such treatments, though they tend to pop-up almost accidentally rather late-on in an analysis text.[1] What is, however, all too often lacking is any systematic *psychological* preparation designed to show up the limitations of our "$q^{th}$ root of the $p^{th}$ power" algorithm, and to prepare us for the unexpected and unexplained restriction to *positive* values of $x$ in the definition

$$x^\alpha = e^{\alpha \ln x} \quad (\text{provided } x > 0)$$

—even though we all think we know perfectly well how to evaluate expressions like $(-1)^n$. This chapter is therefore an attempt to provide some kind of psychological soil in which the logical treatment may eventually take root.

We shall assume that the reader has met ordinary powers and the index laws for positive integer exponents, and that she/he has successfully guessed how to interpret negative and fractional powers, but has not yet woken up to the limitations of the rule

"$x^{p/q}$ equals the $q^{th}$ root of the $p^{th}$ power of $x$."

There are then two obvious sources of possible confusion, whose clarification will provide us with an opportunity to learn something new about functions and something new about the way mathematics itself progresses.

---

[1] For example, in Spivak's *Calculus* (1980) we have to wait until Chapter 17, p. 288; in Hardy's *Pure Mathematics* (1960) we have to wait until Chapter 9, p. 409; in Rudin's *Principles of Mathematical Analysis* (1964) we have to wait until Chapter 8, p. 164.

First of all one cannot really expect the beginner to distinguish between

the *function* $f(x) = x^{1/3}$

and

the *particular algorithm* $f(x) =$ "the cube root of x."

The expression $f(x) = x^{1/3}$ has no meaning on its own and must clearly be explained in terms of something with which we are already familiar. And without the *particular algorithm* ($f(x) =$ "the cube root of x") it is hard to see how the *function* $f(x) = x^{1/3}$ could ever become an object of thought. It simply does not occur to us that there may be *many different algorithms* which would enable us to evaluate *the one function* $f(x) = x^{1/3}$: that, for example, when $0 < x < 2$ we might use the algorithm[2]

$$f(x) = 1 + (x-1) + \frac{1}{3} \cdot \frac{(x-1)^2}{2!} + \frac{1 \cdot -2}{3^2} \cdot \frac{(x-1)^3}{3!}$$

$$+ \frac{1 \cdot -2 \cdot -5}{3^3} \cdot \frac{(x-1)^4}{4!} + \cdots$$

$$= x + \sum_{r=1}^{\infty} \frac{1 \cdot -2 \cdot -5 \cdot \ldots \cdot (4-3r)}{3^r} \cdot \frac{(x-1)^{r+1}}{(r+1)!}.$$

Neither does it occur to us that our particular algorithm may depend on the particular way the function $f(x) = x^{1/3}$ has been written down rather than on the function itself.[3] For example, since

2/6 is exactly the same number as 1/3

we must presumably expect that

$f(x) = x^{2/6}$ is exactly the same function as $f(x) = x^{1/3}$;

[2] In practice, when evaluating an endless sum we will have to make do with an approximate answer; but the same is true however one tries to evaluate "the cube root of x."

[3] In the case of an operation as fundamental as *addition* we do in fact learn both to accept a number of algorithms for addition, and that our choice of algorithm depends on the form in which the numbers are given: consider for example, addition of fractions as in

$$\frac{1}{9} + \frac{2}{7} = \frac{7+18}{63} = \frac{25}{63},$$

addition of ordinary finite decimals as in

$$371.6$$
$$\underline{43.754}$$
$$415.354$$

addition of endless decimals as in Chapter II.11, and "algebraic" addition as in

$$\pi + 2\pi = 3\pi.$$

however our "$q^{\text{th}}$ root of the $p^{\text{th}}$ power" algorithm for "$x^{p/q}$" makes $f(x) = x^{2/6}$ a two-valued function and $f(x) = x^{1/3}$ a single-valued function, and so suggests that they are in fact *different* functions with *different* graphs (Figure 176). More generally, the "$q^{\text{th}}$ root of the $p^{\text{th}}$ power" algorithm for "$x^{p/q}$" suggests that

$$x^{p/q} \neq x^{2p/2q} \text{ when } q \text{ is odd.}$$

Are we then to say that $2p/2q$ can be cancelled down to $p/q$ except when they are being used as exponents?

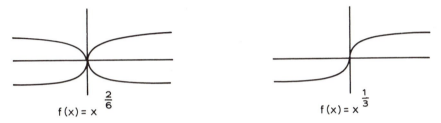

$$f(x) = x^{\frac{2}{6}} \qquad\qquad\qquad f(x) = x^{\frac{1}{3}}$$

Figure 176

But in spite of such indications that the "$q^{\text{th}}$ root of the $p^{\text{th}}$ power" algorithm cannot possibly represent the whole truth, it does, if used carefully, allow us to interpret *rational* powers $x^{p/q}$ in a way which is psychologically compatible with our idea that ordinary integral powers, like $2^2$ and $2^{-2} = (2^{-1})^2$ correspond to repeated multiplication—the operation of taking $q^{\text{th}}$ roots is after all only the inverse of taking $q^{\text{th}}$ powers. Nevertheless, it has been one of the recurring themes of this book, that *new wine eventually bursts old wineskins*: we should not therefore assume that *irrational* (or even complex) powers will automatically fit neatly and tidily into the old conceptual framework of repeated multiplication. In fact though we can extend the definition of the *function $f(x) = x^{\alpha}$* so that it makes sense for irrational exponents $\alpha$, the "$q^{\text{th}}$ root of the $p^{\text{th}}$ power" *algorithm does not generalise.*

A second source of confusion is the remarkable variety of graphs which one obtains for the function $f(x) = x^{p/q}$ as the exponent $p/q$ varies. Thus

$$f(x) = x^{1/3}, \qquad f(x) = x^{3/7}, \qquad f(x) = x^{5/11}, \ldots$$

This natural temptation to identify a function with one particular algorithm by means of which it can be evaluated, and even to insist that all functions should be specified by one particular kind of algorithm, was to some extent responsible for the dominance of the algebraic idea of a function between the middle of the eighteenth and the beginning of the nineteenth century, which was discussed in the previous chapter. The modern function-concept has departed even further from this identification of the function with one particular algorithm by means of which it can be evaluated. Thus Dirichlet's function $\varphi_1$ (page 274) is perfectly well-defined in the sense that $\varphi_1(x)$ is equal to either $c$ or $d$ according as $x$ is rational or irrational, but *there is no algorithm for evaluating $\varphi_1(x)$ in practice*: (for example, as we observed in Chapter IV.1, it is not known whether Euler's constant $\gamma$ is rational or irrational; there is therefore no known way of deciding whether $\varphi_1(\gamma) = c$ or $\varphi_1(\gamma) = d$).

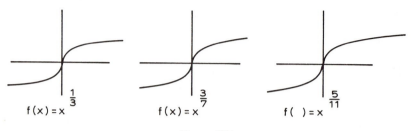

Figure 177

are defined, and single-valued, for all values of $x$, and have graphs which are all more or less the same shape (Figure 177). But between $x^{1/3}$ and $x^{3/7}$ we have $x^{2/5}$, and between $x^{3/7}$ and $x^{5/11}$ we have $x^{4/9}$; and though

$$f(x) = x^{2/5}, \qquad f(x) = x^{4/9}, \qquad f(x) = x^{6/13}, \ldots$$

are defined, and single-valued, for all values of $x$, their graphs are disturbingly different from those of $x^{1/3}$, $x^{3/7}$, $x^{5/11}$ (Figure 178). Moreover, it is clear that both of the sequences

$$x^{1/3}, \quad x^{3/7}, \quad x^{5/11}, \ldots \quad \text{and} \quad x^{2/5}, \quad x^{4/9}, \quad x^{6/13}, \ldots$$

are, in some sense, heading for

$$f(x) = x^{1/2}$$

whose graph is different yet again (Figure 179). The complete picture is even more complicated than this; for between $x^{1/3}$ and $x^{2/5}$ we have $x^{3/8}$, and between $x^{2/5}$ and $x^{3/7}$ we have $x^{5/12}$, etc., and the functions $f(x) = x^{3/8}$ and $f(x) = x^{5/12}$ have graphs which look like that of $f(x) = x^{1/2}$.

There are two standard strategies in mathematics for resolving confusion of this kind:

*Strategy I.* Restrict attention to *what is common* to all known examples, excluding potentially disruptive material in the interests of (temporary) clarity.

*Strategy II.* Look for some "higher" synthesis, or theory, which encompasses all known examples, and which shows up the original confusion as the result of a false, or partially occluded, perspective.

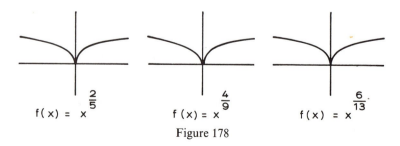

Figure 178

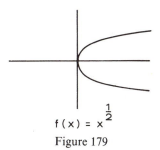

$$f(x) = x^{\frac{1}{2}}$$

Figure 179

Both strategies are important, and each tells us something about the way out of our specific difficulty.

*Strategy I.* In seeking to understand exponents we must be prepared, in the interests of (temporary) clarity, to restrict our attention (temporarily) to *part* of what we think we know.

*Strategy II.* Rather than reject those observations that do not fit into our *temporary* explanation, we should accept them as possible evidence that something more interesting is going on behind the scenes, which may require our attention at a later date.

Implicit in the discussion of the previous chapter was the understanding that a function should be well-defined in the sense that

> "For each particular value given to the variable one can deduce the corresponding value of the function."

We shall (temporarily) interpret this as meaning that "*proper*" *functions must be single-valued*, and shall therefore reject the lower portion of the graph of $f(x) = x^{1/2}$: that is, we shall assume that $x^{1/2}$ denotes the *positive* square root of $x$.

We are then left with three different types of graph (Figure 180). If we decide (Strategy I) to restrict our attention, for the moment, to that which is common to these three types, then we must clearly consider *only positive values of x*. Once this has been done the three graphs become strikingly similar in shape.

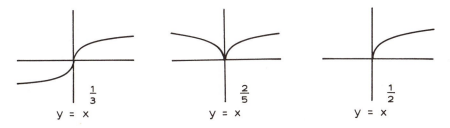

$$y = x^{\frac{1}{3}} \qquad\qquad y = x^{\frac{2}{5}} \qquad\qquad y = x^{\frac{1}{2}}$$

Figure 180

But what if we happen to have graphed $x^0$, $x^{-1/3}$, $x^{-1/2}$ and $x^{-2/3}$ (Figure 181) as well? The same considerations as before lead us to reject the lower portion of the graph of $f(x) = x^{-1/2}$, and suggest that we should (temporarily) restrict our attention to *positive* values of $x$. But this then highlights the natural progression in the graphs from $x^{-2/3}$, through $x^0$ to

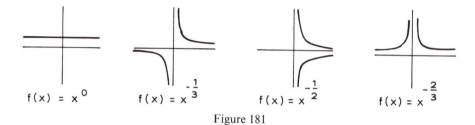

$$f(x) = x^0 \qquad f(x) = x^{-\frac{1}{3}} \qquad f(x) = x^{-\frac{1}{2}} \qquad f(x) = x^{-\frac{2}{3}}$$

Figure 181

$x^{1/2}$ and beyond, and suggests that we should consider all these graphs as sections of the surface

$$y = x^z$$

in 3-dimensions (Figure 182).

Our next task is to guess some suitable meaning for as yet undefined functions such as

$$f(x) = x^{\sqrt{1/2}} = x^{1\sqrt{2}} \quad \text{and} \quad f(x) = x^\pi,$$

at least within the context of our temporary restriction to positive values of $x$. One strategy here is to observe that the curves $y = x^{p/q}$, which may be viewed as sections of the surface $y = x^z$ with $z = p/q$ rational, appear to fit together as part of a single continuous sheet.[4] We would therefore expect that a sensible definition of $x^\pi$ would give rise to a curve $y = x^\pi$ fitting snuggly between the curves $y = x^{3.14}$ and $y = x^{3.15}$ (and even more snugly between the curves $y = x^{3.141}$ and $y = x^{3.142}$). Thus the image of the 3-dimensional surface $y = x^z$ provides us with a plausible *method of calculating approximate values* for $x^z$ when $z$ is irrational, by simply replacing the irrational exponent $z$ by some rational approximation $p/q$. Implicit in this method of calculation is a sort of *geometric meaning* for the function $f(x) = x^z$ when $z$ is irrational, namely,

if $\dfrac{m}{n} < z < \dfrac{p}{q}$ then $f(x)$ is squeezed between $x^{m/n}$ and $x^{p/q}$;

---

[4] At first sight one has the impression that the surface undergoes some sudden change when the value of $z$ passes through 0, but as long as we restrict attention to *positive* values of $x$, this is not, in fact, the case—the only potentially troublesome value being $x = 0$.

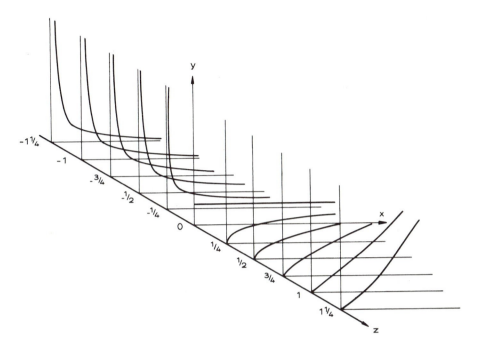

Figure 182

but this does not provide any intuitive algebraic (or analytic) meaning
which we might recognise as a generalisation of the "$q^{th}$ root of the $p^{th}$
power" interpretation of $x^{p/q}$.

The difficulty of attributing intuitive meaning to arbitrary irrational
*powers* is in some ways similar to the difficulty we faced when we first
encountered irrational *numbers*. At that stage, the only numbers we could
prove were irrational were very special numbers like $\sqrt{2}$, but it was perhaps
helpful to consider such examples explicitly however few and however special they may have been. Similarly, there may be some value here in seeking
to interpret certain special *irrational* powers *in the spirit of* the familiar
conceptual framework for *rational* exponents. We may, for example, use the
index laws and a generalised notion of *repeated root extraction* to gain some
feeling for $x^z$ when $z$ is the $n^{th}$ root of a rational number. Observe, for
example, that we may write

$$x^{1/4} \quad \text{as} \quad (x^{1/2})^{1/2};$$

that is, we may interpret the graph of $y = x^{1/4}$ as the result of

first applying the square root ($=\frac{1}{2}^{th}$ power) operation to the curve
$y = x^1$, so "bending" this graph in such a way as to obtain the curve
$y = x^{1/2}$;

then applying the same square root $(=\tfrac{1}{2}^{\text{th}}$ power) operation to the (already bent) curve $y = x^{1/2}$, so "bending" this graph even more to obtain the curve $y = x^{1/4}$.

We may then write

$$x^{1/2} \quad \text{as} \quad (x^{1/\sqrt{2}})^{1/\sqrt{2}}$$

and try to see the graph of $y = x^{1/2}$ as the result of

first applying the "$x^{1/\sqrt{2}}$-operation" to the graph $y = x^1$

to obtain the (unknown) curve $y = x^{1/\sqrt{2}}$, and

then applying the same "$x^{1/\sqrt{2}}$ operation" to the (already bent) curve $y = x^{1/\sqrt{2}}$ to obtain the graph of $y = x^{1/2}$.

One thus expects the curve $y = x^{1/\sqrt{2}}$ to have a curvature somewhere between that of $y = x^1$ and that of $y = x^{1/2}$.

Let us suppose that the fundamental exponential function $e^x$ has already been introduced—as a power series

$$e^x = 1 + x + \frac{x^2}{2!} + \frac{x^3}{3!} + \frac{x^4}{4!} + \dots,$$

or as the number pinpointed by the endless sequence $a_1, a_2, a_3, \dots$, where

$$a_n = \left(1 + \frac{x}{n}\right)^n$$

(see Exercise 1), or as the inverse of the natural logarithm

$$\ln(x) = \int_1^x \frac{1}{t}\, dt,$$

or even as the solution of a particular differential equation

$$\frac{dy}{dx} = y, \qquad y(0) = 1.$$

Then having explored irrational exponents without stumbling on any simple-minded intuitive interpretation, we may even be relieved to meet the *definition*

$$x^z = e^{z \ln(x)} \quad \text{(provided } x > 0).$$

For this certainly agrees with what we already know about $x^{p/q}$, and seems to reflect what we had guessed concerning $x^{1/\sqrt{2}}$, $x^\pi$, etc.: indeed, *if we fix $x$*, then this definition asserts that the corresponding cross-section perpendicular to the $x$-axis of the 3-dimensional surface $y = x^z$ is simply the exponential curve

$$y = e^{z \cdot \ln(x)}.$$

But however satisfactory this definition may be in the context of our temporary restriction to *positive* values of $x$, we shall at some stage have to reconsider, for example,

(i) the fact that the curve $y = x^{1/2}$ has a second (negative) branch;
(ii) the fact that $x^{1/3}$ is defined in a sense that we cannot simply ignore for negative values of $x$;
(iii) the fact that we have already interpreted $(-1)^{1/2} = \pm i$ as a *complex* number;
(iv) the fact that Euler's remarkable formula

$$-1 = e^{i\pi}$$

suggests a way of assigning a value to $(-1)^z$ for arbitrary exponent $z$: namely,

$$(-1)^z = e^{i\pi z}.$$

And once we realise that *complex* numbers are going to crop up even though we are considering *real* numbers, like $-1$, and *real* powers, like $\frac{1}{2}$, then we may notice

that $x^{1/2}$ has *two* values (say $\lambda$ and $-\lambda$) for each value of $x$, positive and negative;

that $x^{1/3}$ has *three* values (say $\lambda$, $\lambda e^{2\pi i/3}$ and $\lambda e^{4\pi i/3}$) for each value of $x \neq 0$;

that $x^{p/q}$ has $q$ values ($\lambda e^{2k\pi i/q}$, $k = 0, 1, \ldots, q - 1$) for each value of $x \neq 0$.

More intriguing still is the fact

that $x^{1/\sqrt{2}}$ has *infinitely many* values ($\lambda e^{2k\pi i/\sqrt{2}}$, where $k$ may be any integer).

You may even begin to wonder why we insisted that $x^z$ should be a *single-valued* function when $x$, $z$ and $x^z$ are all real! The reason is simply that single-valued functions are much easier to work with than multiple-valued functions: the value $f(x)$ of a single-valued function at the point $x$ is completely unambiguous; if the single-valued function $f$ can be differentiated at the point $x$, its derivative $f'(x)$ at that point is unique; when single-valued functions are combined by addition, multiplication, composition, etc., they give rise to other single-valued functions; and so on. Admittedly, the inverse of a single-valued function may *not* be single-valued; but the advantages of *making* it single-valued are so great that it has become standard practice to do this, even though it means that a given function $f$ may give rise to more than one single-valued inverse.

However, our multiple-valued complex exponentials will not simply go away, and we may as well try to squeeze as much mileage as we can out of our definition

$$\text{``}x^z = e^{z \cdot \ln(x)}\text{''}$$

by seeking to make some sense of it when $x$ and $z$ are arbitrary complex numbers. The definition of $e^x$ as a power series can be used to specify the fundamental exponential function $e^w$ as a function of the *complex* variable $w$:

$$e^w = 1 + w + \frac{w^2}{2!} + \frac{w^3}{3!} + \frac{w^4}{4!} + \dots .$$

However, since $e^{2\pi i} = 1$, it follows that

$$e^w = e^{(w + 2k\pi i)} \quad \text{for each integer } k:$$

the inverse function "ln" is therefore *infinite-valued* (since $\ln(e^w) = w + 2k\pi i$ for each integer $k$). Thus if we try to define arbitrary exponentials $x^z$, for arbitrary complex numbers $x$ and $z$, by means of the expression $x^z = e^{z \cdot \ln(x)}$, then we must expect to find multiple, or even infinitely many, values for $x^z$. One can in fact get round the fact that "ln" is infinite-valued by working instead with a related *single-valued* function "Log": the definition

$$x^z = e^{z \cdot \text{Log}(x)}$$

for arbitrary complex numbers $x$ and $z$ is then *single-valued*, and corresponds to a notion of exponential which is very similar in many ways to the one we started out with at the beginning of this chapter; but

$$x^z \cdot y^z = (x \cdot y)^z$$

no longer holds exactly as it stands, and

$$x^{p/q} = (x^p)^{1/q} = (x^{1/q})^p$$

does not hold even in very simple cases.

EXERCISES

1. In Exercise 1 (ii) of Chapter II.10 we showed that, if

$$a_n = \left(1 + \frac{1}{n}\right)^n$$

then the endless sequence $a_1, a_2, a_3, \dots$ was increasing with each $a_n < 3$: we then appealed to the fundamental property of real numbers to conclude that this endless sequence necessarily pinpoints some real number, which we christened $e$. Now let $x$ be any real number, and put

$$b_n = \left(1 + \frac{x}{n}\right)^n.$$

(i)  If $x > 0$, then we shall show that the endless sequence $b_1, b_2, b_3, \dots$ is increasing and that, for some number $K$, each $b_n < K$: we will then define $e^x$ to be the number pinpointed by this endless increasing sequence.

(ii) If $x = -y < 0$, then we shall show that the endless sequence $b_1, b_2, b_3, \ldots$ pinpoints the number $(e^y)^{-1}$ where $e^y$ is as defined in (i): we will then define $e^x$ to be equal to $(e^y)^{-1}$.

(iii) If $x = 0$, then each $b_n = 1$ so the endless sequence $b_1, b_2, b_3, \ldots$ pinpoints the number 1 and we define $e^0$ to be equal to 1.

(i) Use the method outlined in Exercise 1 (ii) of Chapter II.10 to show that, if $x > 0$, then

(a) the endless sequence $b_1, b_2, b_3, \ldots$ is increasing, and

(b) for each $n$, $b_n \leq 1 + x + \dfrac{x^2}{2!} + \dfrac{x^3}{3!} + \ldots + \dfrac{x^n}{n!}$.

Let $N$ be a fixed integer larger than $x$.

(c) Show that, for each $n$,

$$b_n < \sum_{r=0}^{\infty} \frac{N^r}{r!} < \sum_{r=0}^{N} \frac{N^r}{r!} + \frac{N^{N+1}}{N!} = K.$$

(d) Hence show that the endless sequence $b_1, b_2, b_3, \ldots$ pinpoints some real number.

(ii) Let $x < 0$; then $x = -y$ with $y > 0$.

(a) Show that if $0 < \delta < 1$, then

$$\left(1 - \frac{\delta}{n}\right)^{-n} < (1 - \delta)^{-1}.$$

(b) Use this to prove that, if $n > y^2$, then

$$0 < \left(1 - \frac{y}{n}\right)^{-n} - \left(1 + \frac{y}{n}\right)^{n} < e^y \cdot \frac{y^2}{n - y^2}.$$

(c) Hence conclude that, when $x = -y < 0$, the endless sequence $b_1, b_2, b_3, \ldots$ pinpoints the number $(e^y)^{-1}$.

2. Show that $e^x > 0$ for every $x$. [Hint: Consider $x > 0$ and $x < 0$ separately.]

3. If $x < y$, show that $e^x < e^y$. [Hint: Show that, if $0 \leq x < y$, then $(1 + (y/n))^n - (1 + (x/n))^n \geq (1 + y) - (1 + x) = y - x$; hence conclude that $e^y - e^x \geq y - x > 0$. Then consider $x < 0 < y$ and $x < y \leq 0$.]

4. Show that, for each positive integer $N$, there exist real numbers $x$ and $y$ such that

$$N < e^x \quad \text{and} \quad \frac{1}{N} > e^y \ (> 0).$$

Since the function $x \mapsto e^x$ is strictly increasing (that is, $x < y$ implies $e^x < e^y$), it has a well-defined inverse function

$$\ln : e^x \mapsto x.$$

The next task is to show that, for each real number $y > 0$, there exists a real number $x$ such that

$$e^x = y:$$

in other words, we should show that the function $e^x$ *takes every possible positive value* as $x$ runs through all the real numbers. It is a consequence of Exercises 2, 3 and 4 that this is equivalent to showing that *the function $e^x$ is continuous at each point $x_0$*.

If we take this on trust, then for every *positive* real number $y$ and each real number $z$, we can define

$$y^z = e^{(z \cdot \ln y)}.$$

5. Use this definition to prove that, for any two real numbers $x$ and $z$

$$(e^x)^z = e^{(x \cdot z)}.$$

[Hint: $(e^x)^z = y^z$, where $y = e^x$.] Hence show that, for each $y > 0$,

$$(y^x)^z = y^{(x \cdot z)}.$$

6. Let $K$ be any positive real number, and put $a_n = \sqrt[n]{K}$.
   (i) If $(0 <)K < 1$, show that the endless sequence

$$a_1, \quad a_2, \quad a_3, \ldots$$

pinpoints the number 1. [Hint: Use the fundamental property of real numbers to show that it pinpoints some number $\alpha \le 1$. Then show $\alpha < 1$ is impossible.]

(ii) If $K > 1$, show that the endless sequence $a_1, a_2, a_3, \ldots$ pinpoints the number 1.

7. Let $L$ be any real number, and put $b_n = (1 + (L/n^2))^n$. Show that the endless sequence

$$b_1, \quad b_2, \quad b_3, \ldots$$

pinpoints the number 1. [Hint: If $L > 0$, use the fact that $1 < (1 + (L/n^2))^{n^2} < e^L$ and Exercise 6(ii). If $L < 0$, put $M = -L > 0$ and observe that

$$\left(1 - \frac{M}{n^2}\right)^n = \left(1 + \frac{\sqrt{M}}{n}\right)^n \cdot \left(1 - \frac{\sqrt{M}}{n}\right)^n;$$

then apply Exercise 1(i) and (ii).]

8. (i) Let $a$ and $b$ be any two real numbers. Show that, if $n > 2|b|$, then $n + b$ lies between $\frac{1}{2}n$ and $2n$, and that $(1 + (a/n(n + b)))$ lies between $(1 + (a/2n^2))$ and $(1 + (2a/n^2))$.

(ii) Use (i) to show that, if $c_n = (1 + (a/n(n + b)))^n$, then the endless sequence $c_1, c_2, c_3, \ldots$ pinpoints the number 1.

(iii) Use (ii) to show that, if $d_n = (1 + (x/n))^n \cdot (1 + (y/n))^n$, then the endless sequence $d_1, d_2, d_3, \ldots$ pinpoints the number $e^{x+y}$. Hence deduce that

$$e^x \cdot e^y = e^{x+y}.$$

$$\left[\text{Hint:} \left(1 + \frac{x}{n}\right)^n \cdot \left(1 + \frac{y}{n}\right)^n = \left(1 + \frac{(x+y)}{n} + \frac{xy}{n^2}\right)^n\right.$$

$$\left. = \left(1 + \frac{(x+y)}{n}\right)^n \left(1 + \frac{xy}{n(n+x+y)}\right)^n.\right]$$

9. We can now show that the function $e^x$ is continuous at each point $x_0$. Let $x_0$ be a given real number.
   (i) Let $\delta > 0$ and show that

   if     $x_0 - \delta < x < x_0 + \delta$,     then     $|e^x - e^{x_0}| < e^{x_0} \cdot (e^\delta - 1)$.

   (ii) If $1 > \delta > 0$, show that, for each $n \geq 1$

   $$\left(1 + \frac{\delta}{n}\right)^n < \left(1 - \frac{\delta}{n}\right)^{-n}.$$

   Hence show that

   $$0 < e^\delta - 1 \leq \frac{\delta}{1 - \delta}.$$

   [Hint: Use Exercise 1(ii)(a).]
   (iii) Given $\varepsilon > 0$, choose $\delta = \varepsilon/(e^{x_0} + \varepsilon)$ and show that

   if     $x_0 - \delta < x < x_0 + \delta$,     then     $|e^x - e^{x_0}| < \varepsilon$.

10. Work out $(d/dx)(x^{xx})$.

11. Observe that

    $$2 < 3, \quad \text{and} \quad 2^3 < 3^2,$$

    whereas

    $$3 < 4, \quad \text{and} \quad 3^4 > 4^3.$$

    (i) On one pair of axes sketch the two graphs

    $$f(x) = 2^x \quad \text{and} \quad g(x) = x^2.$$

    Show that if $x > 4$, then $2^x > x^2$.
    (ii) On one pair of axes sketch the two graphs

    $$f(x) = 3^x \quad \text{and} \quad g(x) = x^3.$$

    Show that if $x > 3$, then $3^x > x^3$.
    (iii) Decide which is greater: $\pi^e$ or $e^\pi$; then prove your guess is correct. On one pair of axes sketch the two graphs

    $$f(x) = e^x \quad \text{and} \quad g(x) = x^e.$$

    (iv) For each value of $a > 0$, decide precisely when $a^x > x^a$. [Hint: One approach is to compare first (and, if necessary, second) derivatives of $f(x)$ and $g(x)$. One does not, however, have to use calculus: another solution uses only the two facts

$$x > 0 \Rightarrow e^x > 1 + x,$$

and

$$x > a \geq e \Rightarrow \left(\frac{x}{a} - 1\right) \ln a \geq \frac{x}{a} - 1 > 0;$$

—see Problem 26 in *Mathematical Morsels* by Ross Honsberger (published by the Mathematical Association of America (1978)).]

# Index